THE

UNIVERSAL MODERN CAMBIST,

AND

FOREIGN AND DOMESTIC

COMMERCIAL CALCULATOR;

OR,

A DICTIONARY OF NUMERICAL, ARITHMETICAL, AND MATHEMATICAL FACTS, TABLES, DATA, FORMULAS, AND PRACTICAL RULES FOR BUSINESS-MEN, MERCHANTS, BANKERS, BROKERS, AND ACCOUNTANTS.

BY

E. S. WINSLOW,

Author of "Comprehensive Mathematics," "Computists' Manual," "Machinists' and Mechanics' Practical Calculator and Guide," "Tin-plate and Sheet-iron Workers' Monitor."

SEVENTH EDITION, REVISED AND ENLARGED.

BOSTON:
PUBLISHED BY THE AUTHOR.
1872.

FOR LOCAL AGENTS, SEE NEXT PAGE.

Single copies of Winslow's Mathematical Works are delivered at the counter or sent by mail, post-paid, on receipt of price, as follows, viz:

See Advertisement, page iv.

Stereotyped by C. J. Peters and Son,
5 Washington Street.

PREFACE

TO THE COMPREHENSIVE MATHEMATICS.

On presenting this work to the public, it may be proper to state that it has been designed and written mainly for the practical man. It contains a vast array of Numerical, Arithmetical, and Mathematical facts, tables, data, formulas, and rules, pertaining to a great variety of subjects, and applicable to a diversity of ends, as well as much information of a more general nature, valuable to the artisan, and commercial classes; thus meeting the wants, in an eminent degree, of the lovers of the exact sciences, and the practical wants of students in the mathematics.

The facts and data alluded to have been gathered, with much care and patience, from a great variety of sources, or derived, often by toilsome investigations, from known and accredited truths. The care that has been taken in respect to these, it is thought, should secure for this particular department reliance and trust.

The tables, which are numerous, have, with few exceptions, been composed and arranged expressly for the work, and a confidence is felt that they may be relied on for accuracy.

From the valuable works of Dr. Ure, Adcock, Gregory, Grier, Brunton; from the publications of the transactions of London, Edinburgh, and Dublin Philosophical Societies; and from the publications by the Smithsonian Institute, much valuable information has been gained, relating mainly to machinery and the arts; and to these sources the author feels indebted.

The conciseness with which the work has been generally written would, perhaps, be found an objection, were it not that all the propositions and problems of intricacy are accompanied with examples and illustrations, and, in the matters of Geometry, additionally accompanied with diagrams. The whole, it is thought, will appear clear to him who consults it. A prominent feature in the design has been to produce a useful work, and one which in the way of price shall be readily accessible to all.

PREFACE

TO THE UNIVERSAL MODERN CAMBYST, AND FOREIGN AND DOMESTIC COMMERCIAL CALCULATOR.

This work is composed of the first five sections of the author's "COMPREHENSIVE MATHEMATICS." It was thought advisable to publish this portion of that work in a separate form on account of price; more especially as it contains all of a commercial nature treated of in that work. Indeed, the contents of that work were arranged expressly to this end. The Table of Contents in both works is the same: the work being stereotyped, this could not well be avoided. The Table of Contents, therefore, in either work, is that of the "COMPREHENSIVE MATHEMATICS;" and the first five sections thereof, that is, Section I., Section II., Section III., Section A., and Section B., is that of the "Universal Modern Cambyst, and Foreign and Domestic Commercial Calculator."

PREFACE

TO THE TIN-PLATE AND SHEET-IRON WORKERS' MONITOR.

This work is composed of Section VI. of the author's "COMPREHENSIVE MATHEMATICS," with portions of other sections of that work. It embraces all that is contained in the last-mentioned work of special interest to the Tinsmith, as such. It may be relied on for accuracy in all particulars, and is believed to be the first and only reliable work of the kind ever published. It is published in separate form on account of price, and with the view of affording apprentices and students every possible facility of obtaining it. It contains over 100 pages, nearly 50 diagrams, and step-by-step directions for constructing, *mechanically*, not less than 30 unlike and different patterns, embracing all of the more difficult and complicated in use, and several of new and beautiful designs.

CONTENTS.

SECTION A.

SECTION I.

MONEYS OF ACCOUNT, COINS, WEIGHTS, AND MEASURES OF THE UNITED STATES; FOREIGN GOLD COINS, &C.

WEIGHTS AND MEASURES.

SECTION II.

MISCELLANEOUS FACTS, CALCULATIONS, AND MATHEMATICAL DATA.

SECTION III.

PRACTICAL ARITHMETIC.

SECTION IV.

GEOMETRY.

MENSURATION OF LINES AND SUPERFICIES.

SECTION V.

MECHANICAL POWERS, MECHANICAL CENTRES, CIRCULAR MOTION, STRENGTH OF MATERIALS; STEAM, THE STEAM ENGINE, ETC.

SECTION VI.

COVERINGS OF SOLIDS, OR PROBLEMS IN PATTERN CUTTING.

SECTION VII.

SECTION B.

DEFINITIONS

OF THE SIGNS USED IN THE FOLLOWING WORK.

$=$ *Equal to.* The sign of equality; as 16 oz. $=$ 1 lb.
$+$ *Plus*, or *More.* The sign of addition; as $8 + 12 = 20$.
$-$ *Minus*, or *Less.* The sign of subtraction; as $12 - 8 = 4$.
$\times$ *Multiplied by.* The sign of multiplication; as $12 \times 8 = 96$.
$\div$ *Divided by.* The sign of division; as $12 \div 4 = 3$.
$\backsim$ *Difference between the given numbers or quantities;* thus, $12 \backsim 8$, or $8 \backsim 12$, shows that the less number is to be subtracted from the greater, and the difference, or remainder, only, is to be used; so, too, *height* $\backsim$ *breadth*, shows that the difference between the height and breadth is to be taken.
: :: : *Proportion;* as 2 : 4 :: 3 : 6; that is, as 2 *is to* 4, *so is* 3 *to* 6.
$\surd$ *Sign of the square root;* prefixed to any number indicates that the square root of that number is to be taken, or employed; as $\sqrt{64} = 8$.
$\sqrt[3]{}$ *Sign of the cube root;* and indicates that the cube root of the number to which it is prefixed is to be employed, instead of the number itself; as $\sqrt[3]{64} = 4$.
^{2} *To be squared, or the square of;* shows that the square of the number to which it is affixed is the quantity to be employed; as $12^2 \div 6 = 24$; that is, that the square of 12, or $144 \div 6 = 24$.
^{3} Indicates that the cube of the number to which it is subjoined is to to be used; as $4^3 = 64$.
. *Decimal point*, or *separatrix.* See Decimal Fractions.
—— *Vinculum.* Signifies that the two or more quantities over which it is drawn, are to be taken collectively, or as forming one quantity; thus, $\overline{4 + 6} \times 4 = 40$; whereas, without the vinculum, $4 + 6 \times 4 = 28$; also, $12 - \overline{2 \times 3 + 4} = 2$; and $\sqrt{\overline{5^2 - 3^2}} = 4$. So, also, $\sqrt{(5^2 - 3^2)} = 4$, and $(4 + 6) \times 4 = 40$.

$$\frac{4^2}{2} \left\{ \begin{array}{l} \text{half of } 4^2 \text{ or} \\ \text{half of the square of } 4 \end{array} \right\} = 8.$$

$$\left(\frac{4^2}{2}\right)^2 \text{(the square of half the square of 4)} = 64.$$

$\frac{1}{2}b^2$ or $\frac{1}{2}(b)^2$ (half the square of b.)
$(\frac{1}{2}b)^2$ (the square of half b)
$(2b)^2$ (the square of twice b.)

SECTION I.

MONEYS, WEIGHTS AND MEASURES,

OF THE UNITED STATES; — THEIR DENOMINATIONS, VALUES, COMPARATIVE VALUES, MAGNITUDES, &c.

MONEYS OF ACCOUNT OF THE UNITED STATES.

These are the *mill*, the *cent*, the *dime*, and the *dollar*.

10 mills = 1 cent, 10 cents = 1 dime, 10 dimes = 1 dollar.

The *dollar* is the *unit* or ultimate money of account of the United States, or of what is sometimes called *Federal money*.

In practice, the dime, as a denomination of value, is rejected. Thus,

10 mills = 1 cent, and 100 cents = 1 dollar.

This mark, $, is equivalent to the word *dollar*, or *dollars*, in this money.

COINS OF THE UNITED STATES.

Until June, 1834, the government of the United States estimated gold in comparison with silver as 15 to 1, and in comparison with copper as 850 to 1.

From June, 1834, until February, 1853, the same government estimated gold in comparison with silver as 16 to 1, and in comparison with copper as 720 to 1.

For all time since February, 1853, this government has estimated gold in comparison with silver as $14\frac{1026}{1161}$ to 1, and in comparison with copper as 720 to 1.

The standard for mint gold with this government until 1834 was 11 parts pure gold and 1 part alloy, the alloy to consist of silver and copper mixed, not exceeding one half copper.

The gold coins, therefore, struck at the United States mint prior to 1834, are 22 carats fine.

In what, until 1834, constituted a dollar of gold coin of United States mintage, there were put 24.75 grains of pure gold; and 27 grains of the standard mint gold of that day were at that time worth $1. Twenty-seven grains of that gold, or gold of that standard, are now, by the present government standard of valuation, worth $1.0652.

The standard for mint silver with this government until 1834, was 1485 parts pure silver and 179 parts pure copper, = $8\frac{59}{179}$ parts pure silver and 1 part pure copper.

The silver coins, therefore, struck at the United States mint prior to 1834, are $10\frac{295}{416}$ ounces fine.

In that which, until 1834, constituted a dollar of silver coin of this government's mintage, there were put 371¼ grains of pure silver; and 416 grains of the standard mint silver of that day were at that time of the value of $1. Four hundred and sixteen grains of that silver, or silver of that standard, are now, by the present government standard of valuation, worth $1.0744.

The *cent*, until 1834, was of pure copper, and weighed 208 grains; since 1834, pure copper, weight 168 grains.

The standard for mint gold with this government is now, and for all time since June, 1834, has been, 9 parts pure gold and one part alloy, the alloy to consist of silver and copper mixed, not exceeding one half silver.

The gold coins, therefore, struck at the United States mint and dated subsequent to 1834, are $21\frac{3}{5}$ carats fine.

The standard weight for these coins is $25\frac{4}{5}$ grains to the *dollar;* and in every $25\frac{4}{5}$ grains of these coins there are $23\frac{22}{100}$ grains of pure gold.

The standard for mint silver with this government is now, and for all time since June, 1834, has been, 9 parts pure silver and 1 part pure copper.

The silver coins, therefore, struck at the United States mint and dated subsequent to 1834, are $10\frac{4}{5}$ ounces fine.

In what, from June, 1834, until February, 1853, constituted a dollar of silver coin of this government's mintage, there were put 371¼ grains of pure silver; and 412½ grains of the standard mint silver of that day (the present standard) were worth, from June, 1834, until February, 1853, $1. Four hundred twelve and one half grains of this standard of silver are now worth, by the present standard of valuation, $1.0742.

The standard weight for silver coins with this government at present is 384 grains to the *dollar*. (See page *a* 51.)

The foregoing is not applicable to the *silver three-cent pieces*, so called, authorized by the Congress of 1850–51. These pieces are composed of 3 parts silver and 1 part copper; and their standard weight is $12\frac{3}{8}$ grains each. They are worth, at full weight, measured by the present standard of 345.6 grains of fine silver to the dollar, 2.685 cents each.

The *nickel* and *bronze* tokens or coins, authorized by Congress at different times between the years 1856 and 1867, are as follows: —

Nickel 1-*cent* token: 88 parts copper and 12 parts nickel; weight, 72 grains. Nickel 3-*cent* token: copper and nickel mixed, not exceeding one-fourth nickel; weight, 32 grains. Nickel 5-*cent* token: copper and nickel mixed, not exceeding one-fourth nickel; weight, 77.16 grains, or 5 grammes; diameter, 2 centimetres.

Bronze 1-*cent* token: 95 parts copper, and 5 parts tin and zinc; weight, 48 grains. Bronze 2-*cent* token: 95 parts copper, and 5 parts tin and zinc; weight 96 grains.

NOTE. — In the preceding calculations of values, and in the following, as is now the common custom everywhere, no value has been assigned to the alloy, either in the silver coins or the gold coins. In general, it is simply copper, and will not net more, it is assumed, than the cost of recovering. In the United States, the law relating to coinage, previous to 1834, required that the alloy for gold coins should consist of silver and copper mixed, not exceeding one-half copper; and the present law provides that it shall consist of silver and copper mixed, not exceeding one-half silver.

BOSTON, July, 1869.

GOLD, — PURE.

24 carats fine	= pure gold.
1 grain	= $0.043066.
23.22 grains	= $1.00.
1 dwt.	= $1.033592.
1 ounce	= $20.671834.

MINT GOLD, — U. S.

Alloy, practically all copper.

Nine parts of pure gold and one part of alloy, or

21⅗ carats fine	= standard coin.
1 grain	= $0.03876.
25⅘ grains	= $1.00.
1 dwt.	= $0.93023.
1 ounce	= $18.60465.

GOLD COINS, — U. S.

Denominations.	Weight in grains.	Standard value.
Double Eagle, - - - - - -	516	$20.00
Eagle, - - - - - - - -	258	10.00
Half-Eagle, - - - - - - -	129	5.00
Quarter-Eagle, - - - - - -	64½	2.50
Triple Gold Dollar, - - - - -	77⅖	3.00
Gold Dollar, - - - - - -	25⅘	1.00
Eagle, prior to 1834 (com. value = $10.62),	270	10.909
Half do. " " (com. value = $5.31),	135	5.454

Private and Uncurrent.			Weight in Grains.	Sales.
A. Bechtler, N. C.,	$5	piece,		$4.75
" "	$2\frac{1}{2}$	"		2.37
" "	1	"		.93
T. Reed, Georgia,	5	"		4.75
" "	$2\frac{1}{2}$	"		2.37
" "	1	"		.93
Moffat, California,	5	"	129	5.00

SILVER,— PURE.

12 ounces fine = Pure Silver.
1 dwt. = $0.06944.
345.6 grains = $1.
1 ounce = $1.38889.

MINT SILVER.— U. S.

Alloy, all copper.

Nine parts pure silver and one part alloy; or
10 oz. 16 dwts. fine = Standard Coin.
1 dwt. = $0.0625.
384 grains = $1.00.
1 ounce = $1.25.

SILVER COINS.— U. S.

	Weight in Grains.	Standard Value.
Dollar, - - - - - - - -	384	$1.00
Half Dollar, - - - - - -	192	.50
Quarter Dollar, - - - - - -	96	.25
Dime, - - - - - - - -	$38\frac{2}{5}$	.10
Half Dime, - - - - - - -	$19\frac{1}{5}$	.05
Three-Cent Piece, $\frac{3}{4}$ silver and $\frac{1}{4}$ copper,	$12\frac{3}{8}$	.03

The copper coins of the United States are the CENT and HALF CENT; they are of pure copper. The weight of the former is 168 grains, and that of the latter, 84 grains.

NOTE.—The silver coins of the United States, issued since February, 1853, are not legal tender in the United States in sums exceeding *five dollars*.

TABLE,

Exhibiting the standard weight and present par value of the silver coins of the United States, of dates subsequent to 1834, *and prior to* 1853.

	Weight in Grains.	Present par value.
Dollar, - - - - - -	$412\frac{1}{2}$	\$1.0742
Half Dollar, - . - - -	$206\frac{1}{4}$	.5371
Quarter Dollar, - - - -	$103\frac{1}{8}$	.2685
Dime, - - - - - -	$41\frac{1}{4}$	.1074
Half Dime, - - - - -	$20\frac{5}{8}$	.0537
Three-Cent Piece, - - - -	$12\frac{3}{8}$	.03

CURRENCIES OF THE DIFFERENT STATES OF THE UNION.

4 Farthings = 1 Penny, 12 Pence = 1 Shilling, 20 Shillings = 1 Pound.

In Massachusetts, Connecticut, Rhode Island, New Hampshire, Vermont, Maine, Kentucky, Indiana, Illinois, Missouri, Virginia, Tennessee, Mississippi, Texas and Florida, 6 shillings = 1 dollar; \$1 = $\frac{3}{10}$ £.

In New York, Ohio and Michigan, 8 shillings = 1 dollar; \$1 = $\frac{2}{5}$ £.

In New Jersey, Pennsylvania, Delaware and Maryland, 7 shillings and 6 pence = 1 dollar; 1 dollar = $\frac{3}{8}$ £.

In North Carolina, 10 shillings = 1 dollar; \$1 = $\frac{1}{2}$ £.

In South Carolina and Georgia, 4 shillings and 8 pence = 1 dollar; \$1 = $\frac{7}{30}$ £.

NOTE.—These *currencies*, so called, are nominal at present in a great measure. The denominations serve in the different States as verbal expressions of value. But they are neither the names of the moneys of account in any of the States, nor are they the national names of any of the real moneys in circulation. All values in money in the United States are legally expressed in *dollars*, *cents*, and *mills*.

THE METRICAL SYSTEM OF WEIGHTS AND MEASURES.

In this system, the METRE is the basis, and is one forty-millionth of the polar circumference of the earth.

The METRE is the principal unit measure of length; the ARE of *surface;* the STERE of *solidity;* the LITRE of *capacity;* and the GRAM of *weight.*

The gram is the weight, in a vacuum, of one cubic centimetre of pure water at its maximum density.

The Metre, almost exactly .	=	39.37	U. S.	inches.
The Are (100 square metres)	=	3.95367	"	square rods.
The Stere (a cubic metre) .	=	35.31445	"	cubic feet.
The Litre (a cubic decimetre)	= {	61.023	"	" inches.
		1.05668	"	wine quarts.
The Gram	=	15.43235	"	grains.

The divisions by 10, 100, 1,000, of each of these units, are expressed by the same prefixes, viz., *deci, centi, milli;* and the multiples by 10, 100, 1,000, 10,000, of each, by *deca, hecto, kilo, myria.* The former series were derived from the Latin language, the latter from the Greek.

To illustrate with the metre: —

10 *milli*metres = 1 *centi*metre, 10 centimetres = 1 *deci*metre, 10 decimetres = 1 METRE, 10 METRES = 1 *deca*metre, 10 decametres = 1 *hecto*metre, 10 hectometres = 1 *kilo*metre, 10 kilometres = 1 *myria*metre.

In commerce, the ordinary weight is the kilogram, and 100 kilograms (usually called kilos) = 1 quintal; 10 quintals = 1 millier, or tonneau. The kilogram = 15,432.35 ÷ 7000 = 2.20462 avoirdupois pounds.

In practice, the terms *milliare, deciare, decare, kiliare,* and *myriare* are usually dropped, and

100 centare = 1 are; 100 ares = 1 hectare.

Also the terms *millistere, hectostere, kilistere,* and *myriastere,* are usually rejected, and 100 centisteres = 1 decistere; 10 decisteres = 1 stere; 10 steres = 1 decastere = 353.1445 cubic feet.

1 centiare (square metre)	=	1.19598526 square yards.
1 kilometre . . .	=	0.62137 statute miles.
1 hectare . . .	=	2.471 = U. S. acres.
1 kilolitre . . .	=	1 stere = 61,023.377953 cubic in.

A hectolitre = 26.41748 wine gallons = 2.83774 Winchester bush.

NOTE. — The system is the one recommended by the Statistical Congress of 1865 as a general system of weights and measures to be adopted by all nations.

FOREIGN GOLD COINS.

NOTE.—The coins of any country, both gold and silver, circulating as foreign in any other, particularly those of the smaller denominations, are usually held at an estimate below their *standard* par value, compared with the money standard of the country in which they circulate as foreign. Many of them, more particularly the silver, having circulation in the United States, are much worn and otherwise depreciated. In some instances, owing to frequent changes made both with regard to weight and purity, certain of them, having the same name and general appearance, bear a premium at home; others, a discount. Others, again, can hardly be said to have a definable value anywhere. The par value of the old pistole of Geneva, for instance, weighing 103¼ grains, is $3.985, while that of the new, weighing 87¾ grains, would, at the same degree of purity, be worth but $3.386; whereas, owing to its higher standard of fineness, its par value is $3.443. The ducat of Austria, coined in 1831, weighs 53½ grains,—its purity is 23.64, and its par value $2.269; while the half sovereign, closely resembling the ducat, coined in 1835, and weighing 87 grains, has a purity only of 21.64, and a par value, consequently, of but $3.378. The *circulating* value of the ducat in the United States, in general, is $2.20, and that of the half sovereign of Austria, $3.25.

ARGENTINE REPUBLIC.	Standard of purity in carats.	Standard weight in grains.	Par value in Federal money.	Circulating value in Federal money.	Par value per grain. *cts.*
Doubloon to 1832,	19.56	418	$14.671	$	3.50
" to "	20.83	415	15.512		3.73
AUSTRIA.					
Sovereign, half in proportion, to 1785,	22.00	170	6.711	6.50	3.94
Sovereign, half in proportion, since 1785,	21.64	174	6.756	6.50	3.88
Ducat, double in proportion,	23.64	53½	2.269	2.20	4.24
BELGIUM.					
Sovereign, half in pro.,	22.00	170	6.711		3.94

	Standard of purity in carats.	Standard weight in grains.	Par value in Federal money.	Circulating value in Federal money.	Par value per grain. cts.
Twenty Franc, more in pro.	21.50	99½	$3.840	$3.83	3.85
Ducat,				2.20	
BOLIVIA, COLOMBIA, CHILI, ECUADOR, PERU, NEW GRENADA, and MEXICO.					
For the modern coins, &c., of these States, see *Foreign Moneys of Account*, SEC. A.					
Doubloon, (8 E)	20.86	417	15.620	15.60	3.74
Half do.	"	208½	7.810	7.50	"
Quarter do.	"	104¼	3.905	3.75	"
Eighth do.	"	52	1.952	1.75	"
Sixteenth do.	"	26	.976	.90	"
Pistole, half in pro.,				3.75	
BRAZIL.					
For the modern coinage of this Empire, see *Foreign Moneys of Account and Coins*, SEC. A.					
Dobraon,	22.00	828	32.719	32.00	3.95
Dobra,	"	438	17.306	17.00	"
Joannes, (*standard variable*)	"	432	17.064	$13 to $17	"
Half do. do. do.	"	216	8.532	$6 to 8.50	"
Moidore, (BBBB) half in pro., (*standard variable*)	21.79	165	6.451	6.00	3.90
Crusado, do. do.	"	16¼	.635		"
DENMARK.					
Christian d'or	21.74	103	4.018		3.90
Ducat, species,	23.48	53½	2.254	2.20	4.21
" current,	21.03	48	1.811		3.77
FRANCE.					
There are but few gold coins of France now in circulation other than multiples of the standard franc, napoleons, fractional and double, in pro.					
Chr. d'or, double in pro.,	21.60	101	3.914	3.90	3.87

	Standard of purity in carats.	Standard weight in grains.	Par value in Federal money.	Circulating value in Federal money.	Par value per grain. cts.
Franc d'or, double in pro.,	21.60	101	$3.914	$3.90	3.87
Louis d'or, " " " to 1786,	21.49	125½	4.840		3.85
Louis d'or, double in pro., since 1786,	21.68	118	4.573	4.50	3.87
Napoleon (20 F.) double &c.	21.60	99½	3.856	3.83	"
GERMANY.					
BADEN.					
Zehn Gulden, 5 in pro.,	21.60	105½	4.088	4.00	3.87
BAVARIA.					
Carolin,	18.49	149¼	4.952		3.32
Ducat, double in pro.,	23.58	53¾	2.275	2.20	4.23
Maximilian,	18.49	100	3.317		3.31
BRUNSWICK.					
Ducat,	23.22	53½	2.220		4.16
Pistole, double in pro.,	21.60	117¼	4.548		3.87
Ten Thaler, 5 in pro., to 1813,	21.55	202	7.811	7.80	3.86
Ten Thaler, less in pro., since 1813,	21.50	204	7.873	7.80	3.85
HANOVER.					
Ducat, double in pro.,	23.83	53½	2.287	2.20	4.27
George d'or, " " "	21.67	102½	3.987		3.88
Zehn Thaler, 5 " "	21.36	204½	7.838	7.80	3.83
HESSE.					
Ten Thaler, 5 in pro., to 1785,	21.36	202	7.742		"
Ten Thaler, 5 in pro., since 1785,	21.41	203	7.799		3.84
SAXONY.					
Ducat,	23.49	53½	2.256	2.20	4.21
Augustus d'or, double in pro., since 1784.	21.55	102½	3.964		3.86
WURTEMBURG.					
Carolin,	18.51	147½	4.899		3.32
Ducat,	23.28	53½	2.235		4.17

	Standard of purity in carats.	Standard weight in grains.	Par value in Federal money.	Circulating value in Federal money.	Par value per grain. *cts.*
GREAT BRITAIN.					
(*Alloy, since* 1826, *all copper.*)					
The modern gold coins of this Kingdom are the *sovereign*, fractional, double, &c.					
Guinea, half in pro., to 1785,	22.00	127	$5.016		3.95
Guinea, half in pro., since 1785,	"	129½	5.111	$5.00	"
Sovereign, half in pro.,	"	123¼	4.866	4.83	"
Five do.	"	616¼	24.332	24.20	"
Sovereign, (*dragon*) half in pro.,	"	122½	4.838	4.80	"
Double Sovereign (*dragon*)	"	246	9.717	9.67	"
GREECE.					
Twenty Drachm, more in pro.,	21.60	89	3.441	3.40	3.87
HOLLAND.					
Ducat,	23.58	53½	2.263	2.20	4.23
Ryder,	22.00	153	6.043		3.95
Double do.	"	309	12.205		"
Ten Gulden, 5 in pro.,	21.60	103	3.988	3.98	3.87
INDIA.					
Pagoda, star,	19.00	52¾	1.798		3.40
Mohur, (E. I. Co.) 1835.	22.00	180	7.106	6.75	3.95
Half Sovereign, do.				2.41	
BOMBAY.					
Rupee,	22.00	179	7.095		3.96
MADRAS.					
Rupee,	22.00	180	7.106		3.95
ITALY.					
ETURIA, Ruspone,	23.97	161¼	6.935		4.30
GENOA, Sequin,	23.86	53½	2.291		4.28
MILAN, Pistole,	21.76	97½	3.807		3.90
" Sequin,	23.76	53½	2.281		4.26

	Standard of purity in carats.	Standard weight in grains.	Par value in Federal money.	Circulating value in Federal money.	Par value per grain. cts.
MILAN, Twenty Lire, more in proportion,	21.58	99½	$3.853	$3.83	3.86
NAPLES, Ducat, multiples in pro.,	21.43	22½	.865		3.84
NAPLES, Oncetta,	23.88	58	2.485		4.28
PARMA, Doppia, to 1786,	21.24	110	4.192		3.81
" Pistole, since 1796,	20.95	110	4.135		3.75
" Twenty Lire,	21.60	99½	3.859	3.83	3.87
PIEDMONT, Carlino, half in pro., since 1785,	21.69	702	27.321		3.89
PIEDMONT, Pistole, half in pro., since 1785,	21.54	140	5.411		3.86
PIEDMONT, Sequin, half in pro., since 1785,	23.64	53¾	2.280		4.23
PIEDMONT, Twenty Lire, more in pro.,	20.00	99¼	3.563	3.50	3.59
ROME, Ten Scudi, 5 in pro.	21.60	267½	10.368		3.87
" Sequin, since 1760,	23.90	52½	2.251		4.28
SARDINIA, Carlino, ½ in pro.,	21.31	247½	9.465		3.82
TUSCANY, Zechino, double in pro.,	23.86	53¾	2.302		4.30
VENICE, Zechino, double in pro.,	23.84	54	2.310		
MALTA.					
Sequin,	23.70	53½	2.275		4.25
Louis d'or, double and demi in pro.,	20.25	128	4.651		3.63
NETHERLANDS.					
Ducat,	23.52	53½	2.257		4.21
Zehn Gulden, 5 in pro.,	21.55	103¾	4.013	4.00	3.86
POLAND.					
Ducat,	23.58	53½	2.264		4.23
PORTUGAL.					
The modern Portuguese gold coins are the coroa of 5000 reis, parts and multiples in proportion. See SEC. A.					

	Standard of purity in carats.	Standard weight in grains.	Par value in Federal money.	Circulating value in Federal money.	Par value per grain, cts.
Dobraon, 24,000 reis,	22.00	828	$32.706	$32.00	3.95
Dobra,	"	438	17.301	17.00	"
Joannes, (*standard variable*)	"	432	17.064	$13 to $17	"
Half " " "	"	216	8.532	$6 to 8.50	"
Moidore, 4000 reis, "				$4¾ to $4½	
Coroa, 5000 "	"	147½	5.83	5.75	"
Milrea,	22.00	19¾	.780		3.95
PRUSSIA.					
Ducat,	23.49	53½	2.255	2.20	4.21
Frederick d'or, double in pro.,	21.60	102½	3.973		3.87
RUSSIA.					
Ducat,	23.64	54	2.291		4.24
Imperial, (10 R.) half in pro., 1801,	23.55	185¼	7.828		4.22
Imperial, (10 R.) half in pro., since 1818,	22.00	201½	7.949	7.90	3.95
SICILY.					
Oncia, double in pro.,	20.39	68½	2.495		3.64
Twenty Lire, more in pro.,	21.60	99½	3.856	3.83	3.87
SPAIN.					
For the *new* standard of coinage, denominations, &c., of this Kingdom, see *Foreign Moneys of Account and Coins*, Sec. A.					
Doubloon (8 S) parts in pro.	21.45	416½	16.031	16.00	3.84
" (8 E) parts as Bolivian, &c.	20.86	417	15.620	15.60	3.74
Pistole, to 1782,	21.48	103	3.970		3.85
" since "	20.93	104	3.906		3.75
Escudo, to 1788,	20.98	52	1.957		3.76
" since "	20.42	52	1.905		3.66
Coronilla " 1800,	20.29	27	983		3.64
SWEDEN.					
Ducat	23.45	53	2.230		4.20

	Standard of purity in carats.	Standard weight in grains.	Par value in Federal money.	Circulating value in Federal money.	Par value per grain. cts.
SWITZERLAND.					
BERNE, Ducat, double in pro.,	23.53	47	$1.984		4.22
BERNE, Pistole,	21.62	117½	4.558		3.88
GENEVA, Pistole,	21.87	87¾	3.443		3.92
" " (old)	21.51	103¼	3.985		3.85
ZURICH, Ducat, double in pro.,	23.50	53½	2.256		4.21
TURKEY.					
Misseir, half in pro. 1820,	15.88	36½	1.040		2.84
Sequin fonducli,	19.25	53	1.830		3.45
Yeermeeblekblek,	22.88	73¾	3.027		4.10

NOTE.—For full and particular specifications regarding modern foreign gold and silver coins, see *Foreign Moneys of Account and coins*, SEC. A.

The standard silver 5-franc piece of France is worth $1.00471 in the silver coins of the United States; but 5 francs in the standard gold coins of France are worth but $0.964725 in the gold coins of the United States.

LONG OR LINEAR MEASURE.—U. S.

STANDARD.—A brass rod, the length of which, at 62° Fahrenheit, is $\frac{36.0000}{39.1393}$ that of a pendulum beating seconds in *vacuo*, at the level of the sea, at the latitude of London, = $\frac{36.0000}{39.1013}$ at 32° Fah., at the gravitation at New York, = the Yard.

6 points	= 1 line.	5½ yards (16½ ft.)	= 1 rod.
12 lines (72 points)	= 1 inch.	40 rods (220 yds.)	= 1 furlong.
12 inches	= 1 foot.	8 fur. (5280 feet)	= 1 stat. mile.
3 feet (36 inches)	= 1 yard.		

SPECIAL, FOR CLOTH.

2¼ inches	= 1 nail.	4 quarters (36 inches)	= 1 yard.
4 nails (9 inches)	= 1 quarter.		

SPECIAL, FOR LAND.

$7\frac{92}{100}$ inches	= 1 link.	100 links (66 feet)	= 1 chain.
25 links	= 1 rod.	80 chains (320 rods)	= 1 s. mile.

ENGINEER'S CHAIN.

10 inches	= 1 link.*
120 links (100 feet)	= 1 chain.

SHOEMAKER'S MEASURE.

No. 1 is $4\frac{1}{8}$ inches in length, and each succeeding number is an addition of $\frac{1}{3}$ of an inch. No. 1 man's size $= 8\frac{1}{24}$ inches.

MISCELLANEOUS.

Hair's breadth	$= \frac{1}{48}$ inch.	Fathom	= 6 feet.
Digit	= 10 lines.	Knot	$= 47\frac{3}{4}$ feet.
Palm	= 3 inches.	Cable's length	= 120 fathoms.
Hand	= 4 "	Geometrical pace	= 4.4 feet.
Span	= 9 "		

12 particular things	= 1 dozen.
12 dozen (144)	= 1 gross.
12 gross (1728)	= 1 great gross.
20 particular things	= 1 score.
24 sheets of paper	= 1 quire.
20 quires	= 1 ream.

SQUARE OR SUPERFICIAL MEASURE.

(*Length × breadth.*)

144	square	inches	= 1	square	foot.
9	"	feet	= 1	"	yard.
$30\frac{1}{4}$	"	yards	= 1	"	rod.
40	"	rods	= 1		rood.
4		roods	= 1		acre.

SPECIAL, FOR LAND.

$62\frac{454}{625}$	square	inches	= 1	square	link.
10000	"	links	= 1	"	chain.
10	"	chains	= 1	"	acre.

Square rod	=	$272\frac{1}{4}$	square	feet.
Rood	=	1210	"	yards.
		10890	"	feet.
Acre (160 square rods)	=	4840	"	yards.
		43560	"	feet.
Square mile	=	640		acres.
		102400	sq.	rods.

220 × 198	square feet	= 1 acre.
The square of 12.649	" rods	
" " of 69.5701	" yards	
" " of 208.710321	" feet	

CIRCULAR MEASURE.

Minute, or Geographical m. (60″)	= 1.152 s. miles. 6086 feet.	Great Circle	=	360 degrees.
League	= 3 miles.	Equatorial circumference of the earth	=	24897 s. m.
Degree	= 60 geo. miles. 69.158 s. ms.	Equatorial diam.	=	7925 "
Sign ($\frac{1}{12}$ zod.)	= 30 degrees.	Polar diam.	=	7899 "
		Mean radius	=	3955.92 "

NOTE. — In the expressions, *square* feet and *feet* square, there is this difference; viz., the former expresses an area in which there are as many square feet as the *number* named, and the latter an area in which there are as many square feet as the *square* of the number named. The former particularizes no *form* of area, the latter asserts a *square* form.

CUBIC OR SOLID MEASURE. — U. S.

(*Length* × *breadth* × *depth.*)

Cubic foot, 1728 cu. inches = { 1.273 cylindrical feet. / 2200 " inches. / 3300 spherical " / 6600 conical "

Cylindrical foot 1728 " inches = { 0.785398 cubic feet. / 1357.2 " inches. / 2592 spherical " / 5184 conical

27 cubic feet = 1 cubic yard.
40 " of round timber = 1 ton.
42 " of shipping " = 1 ton.
50 " of hewn " = 1 ton.
128 " = 1 cord.

Cubic foot of pure water, at the maximum density at the level of the sea, (39°.83, barometer 30 inches) = { 62½ avoirdupois pounds. / 1000 " ounces.

Cylindrical foot = { 49.1 " pounds. / 785.4 " ounces.

Cubic inch = { 0.036169" pounds / 0.5787 " ounces. / 253.1829 grains.

Cylindrical inch = { 0.028415 avd. pounds. / 0.4546 " ounces

Pound = 27.648 cubic inches.
" distilled = 27.7015 " "
Cubic inch " = 252.6934 grains.
Pound at 62°, distilled = 27.7274 cub. inches.
Cubic inch at 62°, " = 252.458 grains.
" " 39°.83, in *vacuo* = 253.0864 "

Cubic foot of salt water (sea) weighs 64.3 pounds.

GENERAL MEASURE OF WEIGHT. — U. S.

AVOIRDUPOIS.

STANDARD. — The pound is the weight, taken in air, of 27.7015 cubic inches of distilled water at its maximum density, (39°.83 F., the barometer being at 30 inches) = 27.7274 cubic inches of distilled water at 62° = 7000 Troy grains.

$27\frac{11}{32}$ grains = 1 dram.
16 drams ($437\frac{1}{2}$ grs.) = 1 ounce.
16 ounces (7000 grs.) = 1 pound.

SPECIAL — GROSS.

28 pounds = 1 quarter.
4 quarters } = { 1 quintal.
112 pounds } = { 1 cwt.
20 cwt. = 1 ton.

SPECIAL — DIAMOND.

16 parts = 1 grain = 0.8 troy gr.
4 grs. = 1 carat = 3.2 " "

SPECIAL — TROY.

(Exclusively for gold and silver bullion, precious stones, and gold, silver and copper coins, and with reference to their monetary value only.)

24 grains = 1 pennyw't.
20 dwts. (480 grs.) = 1 ounce.
12 oz. (5760 grs.) = 1 pound.

SPECIAL — APOTHECARIES'.

(Exclusively for compounding medicines, for recipes and prescriptions.)

20 grains = 1 scruple, ℈
3 scruples = 1 dram, ʒ
8 drams(480 g.) = 1 ounce, ℥.
12 oz. (5760 g.) = 1 pound, ℔

1 lb. avoir. = $1\frac{31}{144}$ lbs. troy.
1 lb. troy = $\frac{144}{175}$ lbs. avoir.
1 oz. avoir. = $\frac{175}{192}$ oz. troy.
1 oz. troy = $1\frac{17}{175}$ oz. avoir.

NOTE. — The comparative value of diamonds of the same quality is as the square of their respective weights. A diamond of fair quality, weighing 1 carat in the rough state, is estimated worth about $\$9\frac{50}{100}$; and it will require one of twice that weight to make one when worked down equal to 1 carat in weight. Hence, to determine the value of a wrought diamond of any given number of carats: — *Rule*. — Double the weight in carats and multiply the square by 9.50. Thus, the value of a wrought diamond, weighing 2 carats, is $2 + 2 = 4 \times 4 = 16 \times 9.50 = \$152$.

LIQUID MEASURE. — U. S.

The "Wine" or "Winchester" Gallon, of 231 cubic inches capacity, is the Government or Customs gallon of the United States for all liquids, and the legal gallon of each state in which no law exists fixing a state or statute gallon of its own. It contains $58372\frac{1}{5}$ grains of distilled water at 39°.83, the barometer being at 30 inches.

4 gills = 1 pint, 2 pints = 1 quart.
4 quarts, or 231 cubic in. } = { 1 gallon.
0.13368 cub. ft., 294.1176 cyl. in. } = { 8.355 av'd. lbs. pure water.

Liquid gallon of the State of New York,* 281.62 cylindric in.	=	0.128 cubic foot. 221.184 " in. 8 avoid. lbs. pure water at 39°.83, b. 30 in.

Barrel	=	$31\frac{1}{2}$ gallons.	Puncheon	=	84 gallons.
Tierce	=	42 "	Pipe or Butt	=	126 "
Hogshead	=	63 "	Tun	=	252 "

Imperial gallon, 277.274 cub. in.	=	10 av'd lbs. distilled water at 62° F., b. 30 in.
Ale gallon, 282 cub. in.	=	$10\frac{1}{5}$ av'd lbs. pure water at 39°.83, b. 30 in.

1 Wine gallon	=	0.8331 Imperial gallon. 0.8191 Ale " 0.10742 W. bushel.

1 Imperial gallon = 1.2 Wine gallons.

DRY MEASURE. — U. S.

The "Winchester Bushel," so called, of $2150\frac{42}{100}$ cubic inches capacity, is the Government bushel of the United States, and the legal bushel of each state having no special or statute bushel of its own. The standard Winchester bushel measure is a cylindrical vessel having an outside diameter of $19\frac{1}{2}$ inches, an inside diameter of $18\frac{1}{2}$ inches, and an inside depth of 8 inches. The standard "heaped" or "coal" bushel of England was this measure heaped to a true cone 6 inches high, the base being $19\frac{1}{2}$ inches, or equal to the outside diameter of the measure. Its ratio to the even bushel was, therefore, as 1.28, nearly, to 1. The present "Imperial" measure of England has the same outside diameter and the same depth as the Winchester, and an internal diameter of 18.8 inches, and the same height of cone is retained for forming the heaped bushel. Its ratio, therefore, to the even bushel is a trifle less than was that of the Winchester. In the United States the "heaped bushel" is usually estimated at 5 even pecks, or as 1.25 to 1 of the standard even bushel, which, if taken as

* By enactment of the Legislature of the State of New York, this gallon ceased to be the legal gallon of that State, April 11, 1852; and the United States Government gallon, of 231 cubic inches capacity, was adopted in its stead.

the rule, requires a cone on the Winchester measure of 5.4 inches to equal the heaped Winchester bushel.

4 gills . . .	=	1 pint.
2 pints . . .	=	1 quart.
4 quarts . .	=	1 gallon or half peck.
8 quarts . .	=	1 peck.
4 pecks 2150.42 cubic in. 1.244456 " ft. 1.5844 cyl. "	=	1 bushel. 2738 cyl. in. 77.7785 av'd lbs. pure water.
Bushel of the State of New York,* 2816.1955 cyl. in.	=	1.28 cubic feet. 2211.84 " in. 80 av'd lbs. pure water.
Bushel of Connecticut,†	=	1.272 cubic feet. 2198 " in. 79.50 av'd lbs. pure water.
Heaped Win. bushel 1.28—even " "	=	2747.7 cubic in. 1.59 cubic ft.
Imperial bushel	=	2218.192 " in.
Chaldron	=	36 Winch. heaped bushels.
1 Winchester bushel	=	0.9694 Imperial bushel. 9.3092 Wine gallons.
1 Imperial bushel	=	1.0315 Winchester bushels.

NOTE. — The Imperial bushel, mentioned above, is the present legal bushel of Great Britain; and the Imperial gallon, mentioned on the preceding page, is the present legal gallon of Great Britain, for all liquids. The gallon for liquids is the same as the gallon for dry measure. Eight Imperial gallons make one bushel. The subdivisions of the gallon and the bushel, and their denominations, are the same as in the Winchester measures. In Great Britain, in addition to the denominations of dry measure used in the United States, the

Strike, ...	= 2 bushels.	Last, ...	= 80 bushels.
Coomb, ...	= 4 "	Sack of corn, ...	= 3 "
Quarter, ...	= 8 "	Bole of corn, ...	= 6 "
Wey or load, ...	= 40 "	Last of gunpowder, . .	= 42 barrels.

* This bushel ceased to be the legal bushel of this State April 11, 1852, and the United States Government bushel, of $2150\frac{42}{100}$ cubic inches capacity, was adopted as the legal bushel in its stead.

† This bushel is now, January, 1852, no longer the legal bushel of this State, and the standard Winchester bushel is adopted in its stead.

SECTION II.

MISCELLANEOUS FACTS, CALCULATIONS, AND PRACTICAL MATHEMATICAL DATA.

SPECIFIC GRAVITIES.

The specific gravity of a body is its weight relative to the weight of an equal bulk of pure water at the maximum density, (39°.83, b. 30 in.) the water being taken as 1., a cubic foot of which weighs 1000 avoirdupois ounces, or 62½ lbs. The specific gravity, therefore, of any body multiplied by 1000, or, which is the same thing, the decimal being carried to three places of figures, or thousands, as in the following TABLES, the whole taken as an integer equals the number of ounces in a cubic foot of the material: multiplied by 62.5, or considered an integer and divided by 16, it equals the number of pounds in a cubic foot; and multiplied by .036169, or taken as an integer and divided by 27648, it equals the decimal fraction of a pound per cubic inch; by which, it is readily seen, the specific gravity of a commodity being known, its weight per any given bulk is easily and accurately ascertained; as, also, its specific gravity, the weight and bulk being known. The weight of any one article relative to that of any other, is as its respective specific gravity to the specific gravity of the other.

METALS.	Specific gravity.		Specific gravity.
Antimony,	6.712	Gold, pure, hammered,	19.546
Arsenic,	5.810	Iridium,	15.363
Bismuth,	9.823	Iron, cast,	7.209
Bronze,	8.700	" wrought,	7.787
Brass, best,	8.504	Lead,	11.352
Copper, cast,	8.788	Mercury, 32°,	13.598
" wire-drawn,	8.878	" 60°,	13.580
Cadmium,	8.604	" —39°,	15.000
Cobalt,	7.700	Manganese,	8.013
Chromium,	5.900	Molybdenum,	8.611
Glucinium,	3.000	Nickel,	8.280
Gold, pure, cast,	19.258	Osmium,	10.000

	Specific gravity.		Specific gravity
Platinum, cast, . .	19.500	Granite, red, . .	2.625
" hammered, .	20.337	" Lockport, .	2.655
" rolled, .	22.069	" Quincy, .	2.652
Potassium, 60°, . .	0.865	" Susquehanna,	2.704
Palladium, . .	11.870	Grindstone, . . .	2.143
Rhodium, . .	11.000	Gypsum, opaque, .	2.168
Silver, pure, cast, .	10.474	Hone, white, . .	2.876
" hammered, .	10.511	Hornblende, . .	3.600
Sodium, . . .	0.970	Ivory, . . .	1.822
Steel, soft, . .	7.836	Jasper, . . .	2.690
" tempered, .	7.818	Limestone, green, .	3.180
Tin, cast, . .	7.291	" white, .	3.156
Tellurium, . .	6.115	Lime, compact, . .	2.720
Tungsten, . .	17.600	" foliated, . .	2.837
Titanium, . .	4.200	" quick, . .	0.804
Uranium, . .	9.000	Loadstone, . . .	4.930
Zinc, cast, . .	6.861	Magnesia, hyd., . .	2.333
		Marble, common, .	2.686
STONES AND EARTHS.		" white Ital. .	2.708
Alabaster, white, .	2.730	" Rutland, Vt., .	2.708
" yellow, .	2.699	" Parian, . .	2.838
Amber, . . .	1.078	Nitre, crude, . .	1.900
Asbestos, starry, .	3.073	Pearl, oriental, . .	2.650
Borax, . . .	1.714	Peat, hard, . .	1.329
Bone, ox, . . .	1.656	Porcelain, China, .	2.385
Brick, . . .	1.900	Porphyra, red, . .	2.766
Chalk, white, . .	2.782	" green, .	2.675
Charcoal, . . .	.441	Quartz, . . .	2.647
" triturated, .	1.380	Rock Crystal, . .	2.654
Cinnabar, . . .	7.786	Ruby, . . .	4.283
Clay, . . .	1.934	Stone, common, . .	2.520
Coal, bitum. avg., .	1.270	" paving, . .	2.416
" anth. " .	1.520	" pumice, . .	0.915
Coral, red, . .	2.700	" rotten, . .	1.981
Earth, loose, . .	1.500	Salt, common, solid, .	2.130
Emery, . . .	4.000	Saltpetre, refined, .	2.090
Feldspar, . .	2.500	Sand, dry, . . .	1.800
Flint, white, . .	2.594	Serpentine, . .	2.430
" black, . .	2.582	Shale, . . .	2.600
Garnet, . . .	4.085	Slate, . . .	2.672
Glass, flint, . .	2.933	Spar, fluor, . .	3.156
" white, . .	2.892	Stalactite, . . .	2.324
" plate, . .	2.710	Talc, black, . .	2.900
" green, . .	2.642	Topaz, . . .	4.011

	Specific gravity.		Specific gravity
SIMPLE SUBSTANCES, *neither metallic nor gaseous.*		Pine, yellow,	.568
		Poplar, white,	.383
Boron,	1.968	Plum,	.785
Bromine,	2.970	Quince,	.705
Carbon,	3.521	Spruce, white,	.551
Iodine,	4.943	Sassafras,	.482
Phosphorus,	1.770	Sycamore,	.604
Selenium,	4.320	Walnut,	.671
Silicon,	1.184	Willow,	.585
Sulphur,	1.990	Yew, Spanish,	.807
		" Dutch,	.788
WOODS, (*dry.*)			
Apple,	0.793	*Highly seasoned Am.*	
Alder,	.800	Ash, white,	.722
Ash,	.760	Beech,	.624
Beech,	.696	Birch,	.526
Birch,	.720	Cedar,	.452
Box, French,	1.328	Cherry,	.606
" Dutch,	.912	Cypress,	.441
Cedar,	.561	Elm,	.600
Cherry,	.715	Fir,	.491
Chestnut,	.610	Hickory, red,	.838
Cocoa,	1.040	Maple, hard,	.560
Cork,	.240	Oak, white, upland,	.687
Cypress,	.644	" James River,	.759
Ebony, American,	1.331	Pine, yellow,	.541
" foreign,	1.290	" pitch,	.536
Elm,	.671	" white,	.473
Fir, yellow,	.657	Poplar, (tulip,)	.587
" white,	.569	Spruce, white,	.465
Hacmetac,	.592		
Hickory, red,	.900	GUMS, FATS, &C.	
Lignum vitæ,	1.333	Asphaltum,	.905 / 1.650
Larch,	.544		
Logwood,	.913	Beeswax,	.965
Mahogany, Spanish, best,	1.065	Butter,	.942
" " com.,	.800	Camphor,	.988
" St. Domingo,	.720	Gamboge,	1.222
Maple, red,	.750	Gunpowder,	.900
Mulberry,	.897	" shaken,	1.000
Oak, live,	1.120	" solid,	1.550 / 1.800
" white,	.785		
Orange,	.705	Gum, Arabic,	1.454
Pear,	.661	" Caoutchouc,	.933
Pine, white,	.554	" Mastic,	1.074

	Specific gravity.
Honey,	1.450
Ice,	.930
Indigo,	1.009
Lard,	.941
Pitch,	1.253
Rosin,	1.100
Spermaceti,	.943
Starch,	1.530
Sugar, dry,	1.606
Tallow,	.938
Tar,	1.257

LIQUIDS.

Acid, acetic,	1.062
" citric,	1.034
" fluoric,	1.060
" nitric,	1.485
" nitrous,	1.420
" sulphuric,	1.846
" muriatic,	1.200
" silicic,	2.660
Alcohol, anhyd.	.794
" 90 °/o	.834
Beer,	1.034
Blood, human,	1.054
Camphene, pure,	.863
Cider, whole,	1.018
Ether, sulph.,	.715
" nitric,	.908
Milk, cow's,	1.032
Molasses, 75 °/o	1.400
Oils, linseed,	.934
" olive,	.917
" rapeseed,	.927
" sassafras,	1.090
" turpentine, com.,	.875
" sperm, pure,	.874
" whale, p'f'd,	.923
Proof spirits,	.925
Vinegar,	1.025
Water, pure,	1.000
" sea,	1.026
" Dead sea,	1.240

	Specific gravity.
Wine, champagne,	.997
" claret,	.994
" port,	.997
" sherry,	.992

ELASTIC FLUIDS:

The measure of which is atmospheric air, at 60°, b. 30 in., its assumed gravity 1; one cubic foot of which weighs 527.04 grains, = .305 of a grain per cubic inch. It is, at this temperature and density, to pure water at the maximum density, as .0012046 to 1, or as 1 to 830.1.

SIMPLE OR ELEMENTARY GASES.

Hydrogen,	.0689
Oxygen,	1.1025
Nitrogen,	.9720
Fluorine,	
Chlorine,	2.470
Carbon, vapor of, (*theoretically*,)	.422

COMPOUND GASES.

Ammoniacal,	.591
Carbonic acid,	1.525
" oxide,	.973
Carbureted hydrogen,	.559
Chloro-carbonic,	3.389
Cyanogen,	1.818
Muriatic acid gas,	1.247
Nitrous acid gas,	3.176
Nitrous oxide gas,	1.040
Olefiant, gas	.978
Phosphureted hydrogen,	1.185
Sulphureted "	1.177
Steam, 212°	.484
Smoke, of wood,	.900
" of coal,	.102
Vapor, of water,	.623
" of alcohol,	1.613
" of spirits turpentine,	5.013

Weight per Bushel (corn Ex. tariff) *of different Grains, Seeds, &c.*

Articles.	lbs.	Articles.	lbs
Barley, (N. E. 47 lbs.) .	48	Hemp seed, ...	40
Beans, ...	60	Oats, ...	32
Buckwheat, . . .	46	Peas, ...	62
Blue-grass seed . .	14	Rye, ...	56
Corn, (N. Y. 58 lbs.) .	56	Salt, T. I., ...	80
Cranberries, . . .		" boiled, ...	56
Clover seed, . . .	60	Timothy seed, . . .	45
Dried Apples, . . .	22	Wheat, ...	60
" Peaches, . .	33	Potatoes, h'p'd, . . .	60
Flax seed, (N. E. 52 lbs.)	56	Malt, ...	38

Weight per Barrel (*Legal or by Usage*) *of different Articles.*

Flour, . . .	196 lbs.	Cider, in Mass., .	32 gals
Boiled Salt, . .	280 "	Soap, . . .	256 lbs.
Beef, . . .	200 "	Raisins, . . .	112 "
Pork, . . .	200 "	Anchovies, . .	30 "
Pickled Fish, . .	200 "	Lime, . . .	220 "
" " in Massachusetts,	30 gls.	Ground Plaster, Hydraulic Cement, .	300 '

A Gallon of Oil weighs	7¾ lbs.
A " " Molasses, standard, (75 per cent.,)	11⅝ "
A " " Linseed Oil (usage, 7½ lbs.) .	7.788 "
A Firkin of Butter, (legal,)	56 "
A Keg of powder, ...	25 "
A Hogshead of Salt is	8 bush
A Perch of Stone = 24¾ cubic feet.	
A Gallon of Alcohol, 90 per cent., weighs .	6.965 lbs.
A " " Proof Spirits, " .	7.732 "
A " " Wine, (average,) " .	8.3 "
A " " Sperm Oil, " .	7.33 "
A " " Whale " p'f'd, " .	7.71 "
A " " Olive " " .	7.66 "
A " " Spirits Turpentine, " .	7.31 "
A " " Camphene, pure, " .	7.21 "

Weight of Coals, &c., broken to the medium size, per Measure of Capacity.

The average weight of Bituminous Coals, broken as above, is about 62 per cent. that of a bulk of equal dimensions in the solid mass, or of the specific gravity of the article; that of Anthracite is about 57 per cent.

Average weight per cubic foot.	lbs.	Average weight per W. Coal bushel.	lbs.
Anthracite, . . .	54	Anthracite, . . .	86
Bituminous, . . .	50	Bituminous, . . .	80
Charcoal, of pine, . .	18.6	Charcoal, hard wood, .	30
" of hard wood, .	19.02	Coke, best, . . .	32

Practical Approximate Weight in Pounds of Various Articles.

Sand, dry, per cubic foot, ...	95
Clay, compact, per cubic foot, . . .	135
Granite, " " " . . .	165
Lime, quick, " " " . . .	50
Marble, " " " . . .	169
Slate, " " " . . .	167
Peat, hard, " " " . . .	83
Seasoned Beech Wood, per cord, . . .	5616
" Yellow Birch Wood, per cord, . .	4736
" Red Maple Wood, " " . .	5040
" " Oak Wood, " " . .	6200
" White Pine Wood, " " . .	4264
" Hickory Wood, " " . .	6960
" Chestnut Wood, " " . .	4880
Meadow Hay, well settled, per cubic foot, $4\frac{1}{2}$ lbs., or 445 cubic feet = 2000 lbs., or $498\frac{4}{10}$ cubic feet = 1 long ton	
Meadow Hay, in large old stacks, per cubic foot,	5
Clover Hay, in settled bulk, " " "	$4\frac{1}{2}$
Corn on Cob, in crib, " " "	22
" shelled, in bin, " " "	45
Wheat, in bin, " " "	48
Oats, in bin, " " "	$25\frac{1}{2}$
Potatoes, in bin, " " "	$38\frac{1}{2}$
Common Brick, $7\frac{3}{4} \times 3\frac{3}{4} \times 2\frac{1}{4}$ in. " M, . .	4500
Front " $8 \times 4\frac{1}{2} \times 2\frac{1}{2}$ in. " " . .	6185

ROPES AND CABLES.

The STRENGTH of cords depends somewhat upon the fineness of the strands; — damp cordage is stronger than dry, and untarred stonger than tarred; but the latter is impervious to water and less elastic.

SILK cords have three times the strength of those of flax of equal circumference, and MANILLA has about half that of hemp.

Ropes made of IRON WIRE are full three times stronger than those of hemp of equal circumference.

White ropes are found to be most durable. The best qualities of hemp are — 1. *pearl gray;* 2. *greenish;* 3. *yellow.* A brown color has less strength.

THE BREAKING WEIGHT of a good hemp rope is 6400 lbs. per square inch, but no cordage may be counted on with safety as capable of sustaining a weight or strain above half that required to break it, and the weight of the rope itself should be included in the estimate.

THE RELIABLE STRENGTH of a good hemp cable, in pounds, is usually estimated as equal to the square of its circumference in inches $\times$ by 120. That of rope $\times$ 200. Thus, a cable of 9 inches in circumference may be relied on as having a sustaining power $= 9 \times 9 \times 120 = 9720$ lbs.

THE WEIGHT, in pounds, of a cable laid rope, per linear foot = the square of its circumference in inches $\times$ .036, very nearly.

The weight, in pounds, of a linear foot of manilla rope = the square of its circumference in inches $\times$ .03, very nearly. Thus, a manilla rope of three inches circumference weighs per linear foot $3 \times 3 \times .03 = \frac{27}{100}$ lbs., $= 3\frac{7}{10}$ feet per lb.

A good hemp rope stretches about $\frac{1}{6}$, and its diameter is diminished about $\frac{1}{5}$ before breaking.

WEIGHT AND STRENGTH OF IRON CHAINS.

Diameter of Wire in Inches.	Weight of 1 Foot of Chain. lbs.	Breaking Weight of Chain. lbs.	Diameter of Wire in Inches.	Weight of 1 Foot of Chain. lbs.	Breaking Weight of Chain. lbs.
$\frac{3}{16}$	0.325	2240	$\frac{5}{8}$	4.217	26880
$\frac{1}{4}$	0.65	4256	$\frac{11}{16}$	4.833	32704
$\frac{5}{16}$	0.967	6720	$\frac{3}{4}$	5.75	38752
$\frac{3}{8}$	1.383	9634	$\frac{13}{16}$	6.667	45696
$\frac{7}{16}$	1.767	13216	$\frac{7}{8}$	7.5	51744
$\frac{1}{2}$	2.633	17248	$\frac{15}{16}$	9.333	58464
$\frac{9}{16}$	3.333	21728	1	10.817	65632

Comparative Weight of Metals, Weight per Measure of Solidity, &c

	Specific Gravity.	Ratio of Comparison	Pounds in a Cubic Foot.	Pounds in a Cubic Inch.
Iron, wrought or rolled .	7.787	1.	486.65	.28163
Cast Iron, . . .	7.209	.9258	450.55	.26073
Steel, soft, rolled, . .	7.836	1.0064	489.75	.28342
Copper, pure, " . .	8.878	1.1401	554.83	.32110
Brass, best, " . .	8.604	1.1050	537.75	.3112
Bronze, gun metal, . .	8.700	1.1173	543.75	.31464
Lead, ...	11.352	1.4579	709.50	.4106

TABLE,

Exhibiting the Weight in pounds of One Foot in Length of Wrought or Rolled Iron of any size, (cross section,) from ⅛ inch to 12 inches.

SQUARE BAR.

Size in Inches.	Weight in Pounds.	Size in Inches.	Weight in Pounds.	Size in Inches.	Weight in Pounds.	Size in Inches.	Weight in Pounds.
⅛	.053	2⅜	19.066	4⅝	72.305	7¾	203.024
¼	.211	2½	21.120	4¾	76.264	8	216.336
⅜	.475	2⅝	23.292	4⅞	80.333	8¼	230.068
½	.845	2¾	25.560	5	84.480	8½	244.220
⅝	1.320	2⅞	27.939	5⅛	88.784	8¾	258.800
¾	1.901	3	30.416	5¼	93.168	9	273.792
⅞	2.588	3⅛	33.010	5⅜	97.657	9¼	289.220
1	3.380	3¼	35.704	5½	102.240	9½	305.056
1⅛	4.278	3⅜	38.503	5⅝	106.953	9¾	321.332
1¼	5.280	3½	41.408	5¾	111.756	10	337.920
1⅜	6.390	3⅝	44.418	5⅞	116.671	10¼	355.136
1½	7.604	3¾	47.534	6	121.664	10½	372.672
1⅝	8.926	3⅞	50.756	6¼	132.040	10¾	390.628
1¾	10.352	4	54.084	6½	142.816	11	408.960
1⅞	11.883	4⅛	57.517	6¾	154.012	11¼	427.812
2	13.520	4¼	61.055	7	165.632	11½	447.024
2⅛	15.263	4⅜	64.700	7¼	177.672	11¾	466.684
2¼	17.112	4½	68.448	7½	190.136	12	486.656

To determine the weight, in pounds, of one foot in length, or of any length, of a bar of any of the following metals of form prescribed, of any size, multiply the weight in pounds, of an equal length of square rolled iron of the same size, (see table of square rolled iron,) if the weight be sought of

Metal	Form	Multiplier
Iron,	Round rolled, by	.7854
Steel,	Square " "	1.0064
"	Round " "	.7904
Cast Iron,	Square bar, "	.9258
" "	Round " "	.7271
Copper,	Square rolled, "	1.1401
"	Round " "	.8954
Brass,	Square " "	1.105
"	Round " "	.8679
Bronze,	Square bar, "	1.1173
"	Round " "	.8775
Lead,	Square " "	1.4579
"	Round " "	1.145

The weight of a bar of any metal, or other substance, of any given length, of a *flat form*, (and any other form may be included in the rule,) is readily obtained by multiplying its cubic contents (feet or inches) by the weight (pounds, ounces, or grains) of a cubic foot or inch of the article sought to be weighed; that is —

Length × *breadth* × *thickness* × *weight per unit of measure.*

For the weight in pounds of a cubic foot or inch of different metals, see "TABLE of weights of metals per measure of solidity, &c."

OR, FOR FLAT OR SQUARE BARS,

Multiply the sectional area in inches by the length in feet, and that product, if the metal be

Wrought Iron, by	3.3795
Cast " "	3.1287
Steel, "	3.4

EXAMPLE. — Required the weight of a bar of steel, whose length is 7 feet, breadth $2\frac{1}{2}$ inches, and thickness $\frac{3}{4}$ of an inch.

$2.5 \times .75 \times 7 \times 3.4 = 44.625$ lbs. *Ans.*

EXAMPLE. — Required the weight of a cast iron beam, whose length is 14 feet, breadth 9 inches, and thickness $1\frac{1}{2}$ inch.

$14 \times 9 \times 1.5 \times 3.1287 = 591.32$ lbs. *Ans.*

TABLE,

Exhibiting the weight in pounds of One Foot in Length of Round Rolled Iron of any diameter, from $\frac{1}{8}$ inch to 12 inches.

Diameter in inches.	Weight in lbs.	Diam. in inches.	Weight in lbs.	Diam. in inches.	Weight in lbs.	Diam. in inches.	Weight in lbs.
$\frac{1}{8}$	.041	$2\frac{3}{8}$	14.975	$4\frac{5}{8}$	56.788	$7\frac{3}{4}$	159.456
$\frac{1}{4}$	.165	$2\frac{1}{2}$	16.688	$4\frac{3}{4}$	59.900	8.	169.856
$\frac{3}{8}$	.373	$2\frac{5}{8}$	18.293	$4\frac{7}{8}$	63.094	$8\frac{1}{4}$	180.696
$\frac{1}{2}$	.663	$2\frac{3}{4}$	20.076	5	66.752	$8\frac{1}{2}$	191.808
$\frac{5}{8}$	1.043	$2\frac{7}{8}$	21.944	$5\frac{1}{8}$	69.731	$8\frac{3}{4}$	203.260
$\frac{3}{4}$	1.493	3	23.888	$5\frac{1}{4}$	73.172	9	215.040
$\frac{7}{8}$	2.032	$3\frac{1}{8}$	25.926	$5\frac{3}{8}$	76.700	$9\frac{1}{4}$	227.152
1	2.654	$3\frac{1}{4}$	28.040	$5\frac{1}{2}$	80.304	$9\frac{1}{2}$	239.600
$1\frac{1}{8}$	3.360	$3\frac{3}{8}$	30.240	$5\frac{5}{8}$	84.001	$9\frac{3}{4}$	252.376
$1\frac{1}{4}$	4.172	$3\frac{1}{2}$	32.512	$5\frac{3}{4}$	87.776	10	266.288
$1\frac{3}{8}$	5.019	$3\frac{5}{8}$	34.886	$5\frac{7}{8}$	91.634	$10\frac{1}{4}$	278.924
$1\frac{1}{2}$	5.972	$3\frac{3}{4}$	37.332	6	95.552	$10\frac{1}{2}$	292.688
$1\frac{5}{8}$	7.010	$3\frac{7}{8}$	39.864	$6\frac{1}{4}$	103.704	$10\frac{3}{4}$	306.800
$1\frac{3}{4}$	8.128	4	42.464	$6\frac{1}{2}$	112.160	11	321.216
$1\frac{7}{8}$	9.333	$4\frac{1}{8}$	45.174	$6\frac{3}{4}$	120.960	$11\frac{1}{4}$	336.004
2	10.616	$4\frac{1}{4}$	47.952	7	130.048	$11\frac{1}{2}$	351.104
$2\frac{1}{8}$	11.988	$4\frac{3}{8}$	50.815	$7\frac{1}{4}$	139.544	$11\frac{3}{4}$	366.536
$2\frac{1}{4}$	13.440	$4\frac{1}{2}$	53.760	$7\frac{1}{2}$	149.328	12	382.208

To find the weight of an equilateral three-sided cast iron prism.

width of side in inches^2 $\times$ 1.354 $\times$ length in feet = weight in lbs.

EXAMPLE. — A three-sided cast iron prism is 14 feet in length, and the width of each side is 6 inches; required the weight of the prism.

$6^2 \times 1.354 \times 14 = 682.4$ lbs. *Ans.*

To find the weight of an equilateral rectangular cast iron prism.

width of side in inches^2 $\times$ 3.128 $\times$ length in feet = weight in lbs.

To find the weight of an equilateral five-sided cast iron prism.

width of side in inches^2 $\times$ 5.381 $\times$ length in feet = weight in lbs.

To find the weight of an equilateral six-sided cast iron prism.

width of side in inches^2 $\times$ 8.128 $\times$ length in feet = weight in lbs.

To find the weight of an equilateral eight-sided cast iron prism

width of side in inches^2 $\times$ 15.1 $\times$ length in feet = weight in lbs.

To find the weight of a cast iron cylinder.

diameter in inches^2 $\times$ 2.457 $\times$ length in feet = weight in lbs.

In a quantity of cast iron weighing 125 lbs., how many cubic inches?

By tabular weight per cubic inch —

$125 \div .26073 = 479.4$ cubic inches. *Ans.*

Or, by tabular weight per cubic foot —

450.55 : 1728 : : 125 : 479.4 cubic inches. *Ans.*

How many cubic inches of copper will weigh as much as 479.4 cubic inches of cast iron?

By tabular weight per cubic inch —

.3211 : .26073 : : 479.4 : 389.27 cubic inches. *Ans.*

Or, by specific gravities —

8.878 : 7.209 : : 479.4 : 389.27 cubic inches. *Ans.*

Or, by tabular ratio of weight —

$$479.4 \times \frac{.9258}{1.1401} = 389.28.$$

A cast iron rectangular weight is to be constructed having a breadth of 4 inches and a thickness of 2 inches, and its weight is to be 18 lbs.; what must be its length?

$$\frac{18}{4 \times 2 \times .26073} = 8.63 \text{ inches. } \textit{Ans.}$$

A cast iron cylinder is to be 2 inches in diameter, and is to weigh 6 lbs.; what must be its length?

.26073 × .7854 = .2047 lb. = weight of 1 cyl. inch, then

$$\frac{6}{2 \times .2047} = 7.327 \text{ inches. } \textit{Ans.}$$

A cast iron cylinder is to weigh 6 lbs., and its length is to be 7.327 inches; what must be its diameter?

$$\sqrt{\left(\frac{6}{7.327 \times .2047}\right)} = 2 \text{ inches. } \textit{Ans.}$$

A cast iron weight, in the form of a *prismoid*, or the *frustrum of a pyramid*, or the *frustrum of a cone*, is to be constructed that will weigh 14 lbs., and the area of one of the bases is to be 16 inches, and that of the other 4 inches; what must be the length of the weight?

$\sqrt{16 \times 4} = 8$ and $\overline{8 + 16 + 4} \div 3 = 9.33$, and $\frac{14}{9.33 \times .26073}$ = 5.75 inches. *Ans.*

NOTE. — For Rules in detail pertaining to the foregoing, see GEOMETRY, *Mensuration of superficies — of solids.*

A model for a piece of casting, made of dry white pine, weighs 7 lbs.; what will the casting weigh, if made of common brass?

By specific gravities —

.554 : 8.604 : : 7 : 108.71 lbs. *Ans.*

NOTE. — As the specific gravity of the substance of which the model is composed must generally remain to some extent uncertain, calculations of this kind can only be relied on as approximate.

TABLE

Exhibiting the Weight of One Foot in Length of Flat, Rolled Iron; Breadth and Thickness in Inches, Weight in Pounds.

Br. and Th. inch.	Wei't. lbs.	Br. and Th. inch.	Wei't. lbs.	Br. and Th. inch.	Wei't. lbs.	Br. and Th. inch.	Wei't. lbs.
½ by ⅛	.211	1¼ by ⅞	3.696	1¾ by ½	2.957	2⅛ by ½	3.591
¼	.422	1	4.224	⅝	3.696	⅝	4.488
⅜	.634	1⅛	4.752	¾	4.435	¾	5.386
⅝ by ⅛	.264	1⅜ by ⅛	.581	⅞	5.175	⅞	6.284
¼	.528	¼	1.161	1	5.914	1	7.181
⅜	.792	⅜	1.742	1⅛	6.653	1⅛	8.079
½	1.056	½	2.323	1¼	7.393	1¼	8.977
¾ by ⅛	.316	⅝	2.904	1⅜	8.132	1⅜	9.874
¼	.633	¾	3.485	1½	8.871	1½	10.772
⅜	.950	⅞	4.066	1⅝	9.610	2¼ by ⅛	.950
½	1.267	1	4.647	1⅞ by ⅛	.792	¼	1.901
⅝	1.584	1⅛	5.228	¼	1.584	⅜	2.851
⅞ by ⅛	.369	1¼	5.808	⅜	2.376	½	3.802
¼	.739	1½ by ⅛	.634	½	3.168	⅝	4.752
⅜	1.108	¼	1.267	⅝	3.960	¾	5.703
½	1.478	⅜	1.901	¾	4.752	⅞	6 653
⅝	1.848	½	2.534	⅞	5.544	1	7.604
¾	2.218	⅝	3.168	1	6.336	1⅛	8.554
1 by ⅛	.422	¾	3.802	1⅛	7.129	1¼	9.505
¼	.845	⅞	4.435	1¼	7.921	1⅜	10.455
⅜	1.267	1	5.069	1⅜	8.713	1½	11.406
½	1.690	1⅛	5.703	1½	9.505	1⅝	12.356
⅝	2.112	1¼	6.337	1⅝	10.297	1¾	13.307
¾	2.534	1⅜	6.970	1¾	11.089	2⅜ by ⅛	1.003
⅞	2.957	1⅝ by ⅛	.686	2 by ⅛	.845	¼	2.006
1⅛ by ⅛	.475	¼	1.373	¼	1.690	⅜	3.010
¼	.950	⅜	2.059	⅜	2.534	½	4.013
⅜	1.425	½	2.746	½	3.379	⅝	5.016
½	1.901	⅝	3.432	⅝	4.224	¾	6.019
⅝	2.376	¾	4.119	¾	5.069	⅞	7.023
¾	2.851	⅞	4.805	⅞	5.914	1	8.026
⅞	3.326	1	5.492	1	6.759	1⅛	9.029
1	3.802	1⅛	6.178	1⅛	7.604	1¼	10.032
1¼ by ⅛	.528	1¼	6.864	1¼	8.449	1⅜	11.036
¼	1.056	1⅜	7.551	1⅜	9.294	1½	12.039
⅜	1.584	1½	8.237	1½	10.138	1⅝	13.042
½	2.112	1¾ by ⅛	.739	2⅛ by ⅛	.898	1¾	14.046
⅝	2.640	¼	1.478	¼	1.795	2	16.052
¾	3.168	⅜	2.218	⅜	2.693	2½ by ⅛	1.056

TABLE. — *Continued.*

Br. and Th. *inch.*	Weight. *lbs.*	Br. and Th. *inch.*	Weight. *lbs.*	Br. and Th. *inch.*	Weight. *lbs.*	Br. and Th. *inch.*	Weight. *lbs.*
2½ by ¼	2.112	2¾ by 1¾	16.264	3¼ by ⅝	6.865	3¾ by 1⅝	20.594
⅜	3.168	1⅞	17.426	¾	8.238	1¾	22.178
½	4.224	2	18.587	⅞	9.610	1⅞	23.762
⅝	5.280	2⅛	19.749	1	10.983	2	25.347
¾	6.336	2¼	20.911	1⅛	12.356	2¼	28.515
⅞	7.393	2⅞ by ⅛	1.214	1¼	13.729	2½	31.683
1	8.449	¼	2.429	1⅜	15.102	2¾	34.851
1⅛	9.505	⅜	3.644	1½	16.475	4 by ⅛	1.690
1¼	10.561	½	4.858	1⅝	17.848	¼	3.379
1⅜	11.617	⅝	6.073	1¾	19.221	½	6.759
1½	12.673	¾	7.287	1⅞	20.594	¾	10.139
1⅝	13.729	⅞	8.502	2	21.967	1	13.518
1¾	14.785	1	9.716	2¼	24.713	1¼	16.898
1⅞	15.841	1⅛	10.931	2½	27.459	1½	20.277
2	16.898	1¼	12.145	3½ by ⅛	1.478	1¾	23.657
2⅝ by ⅛	1.109	1⅜	13.360	¼	2.957	2	27.036
¼	2.218	1½	14.574	⅜	4.436	2¼	30.416
⅜	3.327	1⅝	15.789	½	5.914	2½	33.795
½	4.436	1¾	17.003	⅝	7.393	2¾	37.175
⅝	5.545	1⅞	18.218	¾	8.871	3	40.555
¾	6.653	2	19.432	⅞	10.350	3¼	43.934
⅞	7.762	2⅛	20.647	1	11.828	4¼ by ⅛	1.795
1	8.871	2¼	21.861	1⅛	13.307	¼	3.591
1⅛	9.980	3 by ⅛	1.267	1¼	14.785	½	7.181
1¼	11.089	¼	2.535	1⅜	16.264	¾	10.772
1⅜	12.198	⅜	3.802	1½	17.743	1	14.363
1½	13.307	½	5.069	1⅝	19.221	1¼	17.954
1⅝	14.416	⅝	6.337	1¾	20.700	1½	21.544
1¾	15.525	¾	7.604	1⅞	22.178	1¾	25.135
1⅞	16.634	⅞	8.871	2	23.657	2	28.726
2	17.742	1	10.139	2¼	26.614	2¼	32.317
2⅛	18.851	1⅛	11.406	2½	29.571	2½	35.908
2¾ by ⅛	1.162	1¼	12.673	2¾	32.528	2¾	39.498
¼	2.323	1⅜	13.941	3¾ by ⅛	1.584	3	43.089
⅜	3.485	1½	15.208	¼	3.168	3¼	46.680
½	4.647	1⅝	16.475	⅜	4.752	3½	50.271
⅝	5.808	1¾	17.743	½	6.337	4½ by ¼	3.802
¾	6.970	1⅞	19.010	⅝	7.921	½	7.604
⅞	8.132	2¼	20.277	¾	9.505	¾	11.406
1	9.294	2	22.812	⅞	11.089	1	15.208
1⅛	10.455	2½	25.345	1	12.673	1¼	19.010
1¼	11.617	3¼ by ⅛	1.373	1⅛	14.257	1½	22.812
1⅜	12.779	¼	2.746	1¼	15.842	1¾	26.614
1½	13.940	⅜	4.119	1⅜	17.426	2	30.416
1⅝	15.102	½	5.492	1½	19.010	2¼	34.218

TABLE. — *Continued.*

Br. and Th. *inch.*	Weight. *lbs.*	Br. and Th. *inch.*	Weight. *lbs.*	Br. and Th. *inch.*	Weight. *lbs.*	Br. and Th. *inch.*	Weight. *lbs.*
4½ by 2½	38.020	4¾ by 3	48.158	5¼ by ¾	13.307	5½ by 2	37.175
2¾	41.822	3¼	52.172	1	17.743	2½	46.469
3	45.624	3½	56.185	1¼	22.178	3	55.762
3¼	49.426	5 by ¼	4.224	1½	26.614	5¾ by ¼	4.858
3½	53.228	½	8.449	1¾	31.049	½	9.716
4¾ by ¼	4.013	¾	12.673	2	35.485	¾	14.574
½	8.026	1	16.898	2¼	39.921	1	19.432
¾	12.040	1¼	21.122	2½	44.356	1¼	24.290
1	16.053	1½	25.347	3	53.228	1½	29.146
1¼	20.066	1¾	29.571	5½ by ¼	4.647	1¾	34.007
1½	24.079	2	33.795	½	9.294	2	38.865
1¾	28.092	2¼	38.020	¾	13.941	2¼	43.723
2	32.106	2½	42.244	1	18.587	2½	48.581
2¼	36.119	3	46.469	1¼	23.234	3	58.297
2½	40.132	5¼ by ¼	4.436	1½	27.881	6 by ¼	5.069
2¾	44.145	½	8.871	1¾	32.528		

WEIGHT OF METALS IN PLATE.

The weight of a SQUARE FOOT *one* inch thick of

Malleable Iron . .	= 40.554	lbs.
Com. plate " . .	= 37.761	"
Cast Iron . . .	= 37.546	"
Copper, wrought . .	= 46.240	"
" com. plate . .	= 45.312	"
Brass, plate, com. . .	= 42.812	"
Zinc, cast, pure . .	= 35.734	"
" sheet . . .	= 37.448	"
Lead, cast . . .	= 59.125	"

And for any other thickness, greater or less, it is the same in proportion; thus, a square foot of sheet copper $\frac{1}{16}$ of an inch thick $= 46.24 \div 16 = 2.89$ lbs. And 5 square feet at that thickness $= 2.89 \times 5 = 14.45$ lbs., &c. So, too, 5 square feet at 2½ inches thickness $= 46.24 \times 2.5 \times 5 = 578$ lbs.

THE AMERICAN WIRE GAUGE.

The American Wire Gauge was prepared by Messrs. Brown and Sharp, manufacturers of machinists' tools, Providence, R. I. It is graded upon geometrical principles, is rapidly becoming the standard gauge with manufacturers of wire and plate in the United States, and cannot fail to supersede the use of the Birmingham Gauge in this country.

TABLE

Showing the Linear Measures represented by Nos. American Wire Gauge and Birmingham Wire Gauge, or the values of the Nos. in the United-States Standard Inch.

No.	American Gauge. *Inch.*	Birm. Gauge. *Inch.*	No.	American Gauge. *Inch.*	Birm. Gauge *Inch.*	No.	American Gauge. *Inch.*	Birm. Gauge. *Inch.*	No.	American Gauge. *Inch.*	Birm. Gauge. *Inch.*
0000	.46000	.454	8	.12849	.165	19	.03589	.042	30	.01003	.012
000	.40964	.425	9	.11443	.148	20	.03196	.035	31	.00893	.010
00	.36480	.380	10	.10189	.134	21	.02846	.032	32	.00795	.009
0	.32486	.340	11	.09074	.120	22	.02535	.028	33	.00708	.008
1	.28930	.300	12	.08081	.109	23	.02257	.025	34	.00630	.007
2	.25763	.284	13	.07196	.095	24	.02010	.022	35	.00561	.005
3	.22942	.259	14	.06408	.083	25	.01790	.020	36	.00500	.004
4	.20431	.238	15	.05707	.072	26	.01594	.018	37	.00445	
5	.18194	.220	16	.05082	.065	27	.01419	.016	38	.00396	
6	.16202	.203	17	.04526	.058	28	.01264	.014	39	.00353	
7	.14428	.180	18	.04030	.049	29	.01126	.013	40	.00314	

Thus the DIAMETER or size of No. 4 *wire*, American gauge, is 0.20431 of an inch; Birmingham gauge, 0.238 of an inch: so the THICKNESS of No. 4 *plate*, American gauge, is 0.20431 of an inch; Birmingham gauge, 0.238 of an inch; and so for the other Nos. on the gauges respectively.

TABLE

Showing the Number of Linear Feet in One Pound, Avoirdupois, of Different Kinds of Wire; Sizes or Diameters corresponding to Nos. American Wire-gauge.

No.	Iron. *Feet.*	Copper. *Feet.*	Brass. *Feet.*	No.	Iron. *Feet.*	Copper. *Feet.*	Brass. *Feet.*
0000	1.7834	1.5616	1.6552	19	293.00	256.57	271.94
000	2.2488	1.9692	2.0872	20	396.41	347.12	367.92
00	2.8356	2.4830	2.6318	21	465.83	407.91	432.35
0	3.5757	3.1311	3.3187	22	587.35	514.32	545.13
1	4.5088	3.9482	4.1847	23	740.74	648.63	687.50
2	5.6854	4.9785	5.2768	24	934.03	817.89	866.90
3	7.1695	6.2780	6.6542	25	1177.7	1031.3	1093.0
4	9.0403	7.9162	8.3906	26	1485.0	1300.4	1378.3
5	11.400	9.9825	10.581	27	1872.7	1639.8	1738.1
6	14.375	12.588	13.342	28	2361.4	2067.8	2191.7
7	18.127	15.873	16.824	29	2977.9	2607.6	2763.8
8	22.857	20.015	21.214	30	3754.8	3287.9	3484.9
9	28.819	25.235	26.748	31	4734.2	4145.5	4394.0
10	36.348	31.828	33.735	32	5970.6	5221.2	5541.4
11	45.829	40.131	42.535	33	7528.1	6592.0	6987.0
12	57.790	50.604	53.636	34	9495.6	8314.9	8813.1
13	72.949	63.878	67.706	35	11972	10483	11111
14	91.861	80.439	85.258	36	15094	13217	14009
15	115.86	100.75	107.53	37	19030	16664	17662
16	146.10	127.94	135.60	38	24003	21018	22278
17	184.26	168.35	171.02	39	30266	26503	28091
18	232.34	203.45	215.64	40	38176	33342	35432

NOTE.—In this TABLE the iron and copper employed are supposed to be nearly pure. The specific gravity of the former was taken at 7.774; that of the latter, at 8.878. The specific gravity of the brass was taken at 8.376.

To find the number of feet in a pound of wire of any material not given in the TABLE, *of any size, American gauge, its specific gravity being known.*

RULE. — Multiply the number of feet in a pound of iron wire of the same size by 7.774, and divide the product by the specific gravity of the wire whose length is sought; or ordinarily, for steel wire, multiply the number of feet in a pound of iron wire of the same size by 0.991.

To find the number of feet in a pound of wire of any given No., Birmingham gauge.

RULE. — Multiply the number of feet in a pound of the same kind of wire, same No., American gauge, by the size, American gauge, and divide the product by the size, Birmingham gauge.

EXAMPLE. — In a pound of copper wire No. 16, American gauge, there are 127.94 feet: how many feet are there of the same kind of wire, same No., Birmingham gauge?

$(127.94 \times .05082) \div .065 = 100.03$. *Ans.*

To find the weight of any given length of wire of any given No. or size, American gauge, or the length in any given weight, by help of the foregoing TABLE.

EXAMPLE. — Required the weight of 600 feet of No. 18 iron wire.

$600 \div 232.34 = 2.5822$ lbs. $=$ 2 lbs. $9\frac{1}{3}$ oz., nearly. *Ans.*

EXAMPLE. — Required the length in feet of $2\frac{1}{2}$ lbs. of No. 31 brass wire.

$4394 \times 2.5 = 10985$. *Ans.*

Characteristics of Alloys of Copper and Zinc — Brass.

Parts by Weight.		Specific Gravity.	Color.	Denomination.
Copper.	Zinc.			
83	17	8.415	Yellowish Red.	Bath Metal.
80	20	8.448	" "	Dutch Brass.
$74\frac{1}{2}$	$25\frac{1}{2}$	8.397	Pale yellow.	Rolled Sheet Brass.
66	34	8.299	Full "	English Sheet Brass.
$49\frac{1}{2}$	$50\frac{1}{2}$	8.230	" "	German Sheet Brass.
33	67	8.284	Deep "	Watchmaker's Brass.

NOTE. — To alloys of copper and zinc, generally, there is added a small quantity of lead, which renders them the better adapted for turning, planing, or filing; and, for the same reason, to alloys of copper and tin, there is usually added a small quantity of zinc (see ALLOYS AND COMPOSITIONS).

TABLE

Showing the Weight of One Square Foot of Rolled Metals, thickness corresponding to Nos., American Wire-gauge.

Thickness. *No.*	Iron. *Pounds.*	Steel. *Pounds.*	Copper. *Pounds.*	Brass. *Pounds.*	Lead. *Pounds.*	Zinc. *Pounds.*
1	10.849	10.999	13.109	12.401	17.102	10.833
2	9.6611	9.7953	11.674	11.043	15.228	9.6466
3	8.6032	8.7227	10.396	9.8340	13.562	8.5903
4	7.6616	7.7680	9.2578	8.7576	12.078	7.6501
5	6.8228	6.9175	8.2442	7.7988	10.755	6.8126
6	6.0758	6.1601	7.3416	6.9450	9.5779	6.0667
7	5.4105	5.4856	6.5377	6.1845	8.5291	5.4024
8	4.8184	4.8853	5.8222	5.5077	7.5957	4.8112
9	4.2911	4.3507	5.1851	4.9050	6.7645	4.2847
10	3.8209	3.8740	4.6169	4.3675	6.0233	3.8151
11	3.4028	3.4501	4.1117	3.8896	5.3642	3.3977
12	3.0303	3.0720	3.6616	3.4638	4.7770	3.0257
13	2.6985	2.7360	3.2607	3.0845	4.2539	2.6934
14	2.4035	2.4365	2.9042	2.7473	3.7889	2.3999
15	2.1401	2.1698	2.5829	2.4463	3.3737	2.1369
16	1.9058	1.9322	2.3028	2.1784	3.0043	1.9029
17	1.6971	1.7207	2.0506	1.9399	2.6753	1.6945
18	1.5114	1.5324	1.8263	1.7276	2.3826	1.5091
19	1.3459	1.3646	1.6263	1.5384	2.1217	1.3439
20	1.1985	1.2152	1.4482	1.3700	1.8893	1.1967
21	1.0673	1.0821	1.2897	1.2300	1.6768	1.0657
22	.95051	.96371	1.1485	1.0865	1.4984	.94908
23	.84641	.85815	1.0227	.96749	1.3343	.84514
24	.75375	.76422	.91078	.86158	1.1882	.75262
25	.67125	.68057	.81109	.76728	1.0582	.67024
26	.59775	.60605	.72228	.68326	.94229	.59685
27	.53231	.53970	.64345	.60846	.83913	.53151
28	.47404	.48062	.57280	.54185	.74728	.47333
29	.42214	.42800	.51009	.48242	.66546	.42151
30	.37594	.38116	.45426	.42972	.59263	.37538

NOTE.—In calculating the foregoing TABLE, the specific gravities were taken as follows: viz., iron, 7.200; steel, 7.300; copper, 8.700; brass, 8.230; lead, 11.350; Zinc, 7.189.

TIN PLATES.

Brand Marks.	Size of Sheets in Inches.	No. of Sheets in Box.	Net Weight in lbs.	Brand Marks.	Size of Sheets in Inches.	No. of Sheets in Box.	Net Weight in lbs.
IC	14 × 14	200	140	SDXX	15 × 11	200	210
IC	14 × 10	225	112	SDXXX	15 × 11	200	231
HC	14 × 10	225	119	SDXXXX	15 × 11	200	252
HX	14 × 10	225	147	TT	14 × 10	225	112
IX	14 × 10	225	140	" IC	12 × 12	225	119
IXX	14 × 10	225	161	" IX	12 × 12	225	147
IXXX	14 × 10	225	182	" IXX	12 × 12	225	168
IXXXX	14 × 10	225	203	" IXXX	12 × 12	225	189
IX	14 × 14	200	174	" IXXXX	12 × 12	225	210
IXX	14 × 14	200	200	" IC	20 × 14	112	112
DC	17 × 12½	100	105	" IX	20 × 14	112	140
DX	17 × 12½	100	126	" IXX	20 × 14	112	161
DXX	17 × 12½	100	147	" IXXX	20 × 14	112	182
DXXX	17 × 12½	100	168	" IXXXX	20 × 14	112	203
DXXXX	17 × 12½	100	189	Ternes IC	20 × 14	112	112
SDC	15 × 11	200	168	" IX	20 × 14	112	140
SDX	15 × 11	200	189				

NOTE. — The above TABLE includes all the regular sizes and qualities of tin plates, except "*wasters*." Other sizes, such as 10 × 10, 11 × 11, 13 × 13, &c., of the different brands, are often imported into the United States to order.

Common English Sheet Iron, Nos. 10 to 28, Birmingham gauge, widths from 24 to 36 inches.

R. G. Sheet Iron, Nos. 10 to 30, Birmingham gauge, widths from 24 to 36 inches.

American Puddled Sheet Iron, Nos. 22 to 28, Birmingham gauge, widths from 24 to 36 inches.

Russia Sheet Iron, Nos. 16 to 8 inclusive, Russia gauge, sheets 28 × 56 inches.

Sheet Zinc, Nos. 16 to 8, Liege gauge, widths from 24 to 40 inches; length 84 inches.

Copper Sheathing, 14 × 48 inches, 14 to 32 oz. (even numbers), per square foot.

Yellow Metal, in sheets, 48 × 14 inches, 14 to 32 oz. (even numbers), per square foot.

5

TABLE

Showing the Capacity, in Wine Gallons, of Cylindrical Cans, of different diameters, at One Inch depth. Diameter in Inches.

Diam'r. *inches.*	Gallons.	Diam'r. *inches.*	Gallons.	Diam'r. *inches.*	Gallons.	Diam'r. *inches.*	Gallons.
6	.1224	12¼	.5102	18½	1.164	24¾	2.083
6¼	.1328	12½	.5313	18¾	1.195	25	2.125
6½	.1437	12¾	.5527	19	1.227	25¼	2.167
6¾	.1549	13	.5746	19¼	1.260	25½	2.211
7	.1666	13¼	.5969	19½	1.293	25¾	2.254
7¼	.1787	13½	.6197	19¾	1.326	26	2.298
7½	.1913	13¾	.6428	20	1.360	26¼	2.343
7¾	.2042	14	.6664	20¼	1.394	26½	2.388
8	.2176	14¼	.6904	20½	1.429	26¾	2.433
8¼	.2314	14½	.7149	20¾	1.464	27	2.479
8½	.2457	14¾	.7397	21	1.499	27¼	2.524
8¾	.2603	15	.7650	21¼	1.535	27½	2.571
9	.2754	15¼	.7907	21½	1.572	27¾	2.518
9¼	.2909	15½	.8169	21¾	1.608	28	2.666
9½	.3069	15¾	.8434	22	1.646	28¼	2.713
9¾	.3233	16	.8704	22¼	1.683	28½	2.762
10	.3400	16¼	.8978	22½	1.721	28¾	2.810
10¼	.3572	16½	.9257	22¾	1.760	29	2.859
10½	.3749	16¾	.9539	23	1.799	29¼	2.909
10¾	.3929	17	.9826	23¼	1.837	29¾	3.009
11	.4114	17¼	1.0120	23½	1.877	30	3.060
11¼	.4303	17½	1.0410	23¾	1.918	30½	3.163
11½	.4497	17¾	1.0710	24	1.958	31	3.264
11¾	.4694	18	1.1020	24¼	1.999	31½	3.374
12	.4896	18¼	1.1320	24½	2.041	32	3.482

Applications of the foregoing TABLE.

EXAMPLE. — A cylindrical can is 11¼ inches in diameter, and its depth is 18⅗ inches; required its capacity.

$.4303 \times 18\frac{3}{5} = 8$ gallons. *Ans.*

EXAMPLE. — The diameter of a can containing oil is 26½ inches, and the oil is 14½ inches in depth. How many gallons are there of the oil?

$2.388 \times 14\frac{1}{2} = 34.6$ gallons. *Ans.*

EXAMPLE. — A can is to be constructed that will hold just 36 gallons, and its diameter is to be 18 inches; what must be its depth?

$36 \div 1.102 = 32\frac{2}{3}$ inches. *Ans.*

EXAMPLE. — A cylindrical can is to be constructed that shall have a depth of 15 inches and a capacity of just 5 gallons; what must be its diameter?

$5 \div 15 = .3333$ = capacity of can in gallons for each inch of depth; and against .3333 gallon in the table, or the quantity in gallons nearest thereto, is 10 inches, the required, or nearest tabular diameter. *Ans.*

NOTE. — The table is not intended to meet demands of the nature of the one contained in the last example, with accuracy, unless the fractional part of the diameter, if there be a fractional part, is ¼, ½ or ¾ inch. As, however, the diameter opposite the tabular gallon nearest the one sought, even at its greatest possible remove, can be but about ⅛ inch from the diameter required, we can, by inspection, determine the diameter to be taken, or true answer to the inquiry, sufficiently near for practical purposes, be the fraction what it may. Or, to throw the demand into a mathematical formula: As the tabular gallon nearest the one sought is to the diameter opposite, so is the tabular gallon required to the required diameter, nearly. Thus, in answer to the last query,

.3400 : 10 :: 3333 : 9.8 inches, the required or true diameter, nearly.

For a mathematical formula strictly applicable to this question, see GAUGING

Or, for a formula more strictly geometrical, we have

$$\sqrt{\frac{\text{Capacity} \times 231}{\text{Depth} \times .7854}} = \text{diameter}.$$

The true diameter, therefore, for the supposed can, is

$$\sqrt{\frac{231 \times 5}{15 \times .7854}} = 9.9- \text{ inches}.$$

WEIGHT OF PIPES.

The weight of ONE FOOT IN LENGTH of a pipe, of any diameter and thickness, may be ascertained by multiplying the square of its exterior diameter, in inches, by the weight of 12 *cylindrical* inches of the material of which the pipe is composed, and by multiplying the square of its interior diameter, in inches, by the same factor and subtracting the product of the latter from that of the former, — the remainder or difference will be the weight. This is evident from the fact that the process obtains the weight of two solid cylinders of equal length, (one foot,) the diameter of one being that of the pipe, and the other that of the vacancy, or bore. For very large pipes, the dimensions may be taken in feet, and the weight of a cylindrical foot of the material used as the factor, or multiplier, if desired.

The weight of 12 *cylindrical* inches (length 1 foot, diameter 1 inch) of

Malleable Iron	= 2.6543 lbs.
Cast Iron	= 2.4573 "
Copper, wrought,	= 3.0317 "
Lead "	= 3.8697 "
Cast Iron — 1 *cyl.* foot —	= 353.86 "

Therefore — EXAMPLE. — Required the weight of a copper pipe whose length is 5 feet, exterior diameter 3¼ inches, and interior diameter 3 inches.

$3\frac{1}{4} = \frac{13}{4} \times \frac{13}{4} = 10.5625 \times 3.0317 = 32.022 +$

$3 \times 3 = 9 \times 3.0317 = 27.285 +$

Ans. $4.737 \times 5 = 23.685$ lbs.

EXAMPLE. — Required the weight of a cast iron pipe, whose length is 10 feet, exterior diameter 38 inches, and interior diameter 3 feet.

$38^2 \times 2.4573 - 36^2 \times 2.4573 = 363.68 \times 10 = 3636.8$ lbs. *Ans.*

Or, $\overline{38^2 - 36^2} = 148 \times 2.4573 = 363.68 \times 10 = 3636.8$ lbs. *Ans.*

EXAMPLE. — Required the weight of a lead pipe, whose length is 1200 feet, exterior diameter $\frac{7}{8}$ of an inch, and interior diameter $\frac{9}{16}$ of an inch.

$\frac{7}{8} \times \frac{7}{8} = \frac{49}{64} = .765625$, and $\frac{9}{16} \times \frac{9}{16} = \frac{81}{256} = .316406$, and $.765625 - .316406 = .449219 \times 3.8697 \times 1200 = 2086$ lbs. *Ans.*

EXAMPLE. — The length of a cast-iron cylinder is 1 foot, its exterior diameter is 12 inches, and its interior diameter 10 inches: required its weight.

$12^2 - 10^2 = 44 \times 2.4573 = 108.12$ lbs. *Ans.*

Or, $144 : 353.86 :: 44 : 108.12$ lbs. *Ans.*

The following TABLE *exhibits the coefficients of weight, in pounds, of one foot in length, of various thicknesses, of different kinds of pipe, of any diameter whatever.*

Thickness in Inches.	Wrought Iron.	Copper.	Lead.
$\frac{1}{32}$	.332	.379	.484
$\frac{1}{16}$	.664	.758	.9675
$\frac{3}{32}$	.995	1.137	1.451
$\frac{1}{8}$	1.327	1.516	1.935
$\frac{5}{32}$	1.658	1.894	2.417
$\frac{3}{16}$	1.99	2.274	2.901
$\frac{7}{32}$	2.323	2.653	3.386
$\frac{1}{4}$	2.654	3.032	3.87
$\frac{5}{16}$	3.318	3.79	4.837
$\frac{3}{8}$	3.981	4.548	5.805

CAST IRON.

Thickness.	Factor.	Thickness.	Factor.	Thickness.	Factor.
$\frac{3}{16}$	1.842	$\frac{5}{8}$	6.143	$1\frac{1}{4}$	12.287
$\frac{1}{4}$	2.457	$\frac{3}{4}$	7.372	$1\frac{1}{2}$	14.744
$\frac{3}{8}$	3.686	$\frac{7}{8}$	8.6	$1\frac{3}{4}$	17.201
$\frac{1}{2}$	4.901	1	9.829	2	19.659

To obtain the weight of pipes by means of the above TABLE—

RULE.—Multiply the diameter of the pipe, taken from the interior surface of the metal on the one side to the exterior surface on the opposite, (interior diameter + thickness,) in inches, by the number in the table under the respective metal's name, and opposite the thickness corresponding to that of the pipe—the product will be the weight, in pounds, of ONE foot in length of the pipe, and that product multiplied by the length of the pipe, in feet, will give the weight for any length required.

EXAMPLE.—Required the weight of a copper pipe whose length is 5 feet, interior diameter and thickness $3\frac{1}{8}$ inches, and thickness $\frac{1}{8}$ of an inch.

$$3\frac{1}{8} = \frac{25}{8} = 3.125 \times 1.516 \times 5 = 23.687 \text{ lbs.} \quad \textit{Ans.}$$

EXAMPLE.—Required the weight of a cast iron pipe, 10 feet in length, whose interior diameter is 3 feet, and whose thickness is 1 inch.

$$36 + 1 = 37 \times 9.829 \times 10 = 3636.73 \text{ lbs.} \quad \textit{Ans.}$$

WEIGHT OF CAST IRON AND LEAD BALLS.

To find the weight of a sphere or globe of any material—

RULE.—Multiply the cube of the diameter, in inches, or feet, by the weight of a *spherical* inch or foot of the material.

The weight of a spherical inch of

Cast Iron . = .1365 lbs.
Lead . . = .215 "

Therefore—EXAMPLE.—Required the weight of a leaden ball whose diameter is $\frac{1}{4}$ of an inch.

$\frac{1}{4} \times \frac{1}{4} \times \frac{1}{4} = \frac{1}{64} = .015625 \times .215 = .00336$ lb. *Ans.*

EXAMPLE.—Required the weight of a cast iron ball whose diameter is 8 inches.

$8^3 \times .1365 = 69.888$ lbs. *Ans.*

EXAMPLE.—How many leaden balls, having a diameter $\frac{1}{4}$ of an inch each, are there in a pound?

$1 \div .00336 = \frac{100000}{336} = 298$. *Ans.*

EXAMPLE.—What must be the diameter of a cast iron ball, to weigh 69.888 lbs?

$69.888 \div .1365 = \sqrt[3]{512} = 8$ inches. *Ans.*

EXAMPLE.—What must be the diameter of a leaden ball to equal in weight that of a cast iron ball, whose diameter is 8 inches?

[Lead is to cast iron as .215 to .1365, as 1.575 to 1.]

$8^3 = 512 \div 1.575 = \sqrt[3]{325} = 6.875$ inches. *Ans.*

WEIGHT OF HOLLOW BALLS OR SHELLS.

The weight of a hollow ball is the weight of a solid ball of the same diameter, *less* the weight of a solid ball whose diameter is that of the interior diameter of the shell.

EXAMPLE.—Required the weight of a cast iron shell whose exterior diameter is $6\frac{1}{4}$ inches, and interior diameter $4\frac{1}{4}$ inches.

$6\frac{1}{4} = \frac{25}{4} \times \frac{25}{4} \times \frac{25}{4} = 244.14 \times .1365 = 33.33$
$4\frac{1}{4} = 4\ 25^3 \times .1365 \qquad = 10.48$

22.85 lbs. *Ans.*

Or, If we multiply the difference of the cubes, in inches, of the two diameters—the exterior and interior—by the weight of a spherical inch, we shall obtain the same result.

EXAMPLE.—Required the weight of a cast iron shell whose exterior diameter is 10 inches and interior diameter 8 inches.

$\overline{10^3 - 8^3} \times .1365 = 66.612$ lbs. *Ans.*

ANALYSIS OF COALS.

Description.	Volatile Matter.	Carbon.	Ash.
Breckinridge, Ky.,	62.25	29.10	8.65
"Albert," N. B.,	61.74	32.14	6.12
Chippenville, Pa.,	49.80		
Kanawha, "	41.85		
Pittsburg, "	32.95		
Cannel, sp. gr. 1.4	35.28	64.72	
Newcastle,	24.72	75.28	
Cumberland,	18.40	80.	1.60
Anthracite, a'v'g.,	3.43	89.46	7.11

Woods of most descriptions vary little from 80 per cent. volatile matter, and 20 per cent. charcoal.

TABLE.— *Exhibiting the Weights, Evaporative Powers, &c., of Fuels, from Report of Professor Walter R. Johnson.*

Designation of Fuel.	Specific Grav-ity.	Weight per Cubic Foot.	Lbs. of Water at 212 degrees converted into Steam by 1 Cubic Foot of Fuel.	Lbs. of Water at 212 degrees converted into Steam by 1 lb. of Fuel.	Weight of Clinkers from 100 lbs. of Coal.
ANTHRACITE COALS.					
Beaver Meadow, No. 3	1.610	54.93	526.5	9.21	1.01
Beaver Meadow, No. 5	1.554	56.19	572.9	9.88	.60
Forest Improvement	1.477	53.66	577.3	10.06	.81
Lackawanna	1.421	48.89	493.0	9.79	1.24
Lehigh	1.590	55.32	515.4	8.93	1.08
Peach Mountain	1.464	53.79	581.3	10.11	3.03
BITUMINOUS COALS.					
Blossburgh	1.324	53.05	522.6	9.72	3.40
Cannelton, Ia.	1.273	47.65	360.0	7.34	1.64
Clover Hill	1.285	45.49	359.3	7.67	3.86
Cumberland, *average*,	1.325	53.60	552.8	10.07	3.33
Liverpool	1.262	47.88	411.2	7.84	1.86
Midlothian	1.294	54.04	461.6	8.29	8.82
Newcastle	1.257	50.82	453.9	8.66	3.14
Pictou	1.318	49.25	478.7	8.41	6.13
Pittsburgh	1.252	46.81	384.1	8.20	.94
Scotch	1.519	51.09	369.1	6.95	5.63
Sydney	1.338	47.44	386.1	7.99	2.25
COKE.					
Cumberland		31.57	284.0	8.99	3.55
Midlothian		32.70	282.5	8.63	10.51
Natural Virginia	1.323	46.64	407.9	8.47	5.31
WOOD.					
Dry Pine Wood		21.01	98.6	4.69	

MENSURATION OF LUMBER.

To find the contents of a board.

RULE. — Multiply the length in feet by the width in inches, and divide the product by 12; the quotient will be the contents in square feet.

EXAMPLE. — A board is 16 feet long and 10 inches wide; how many square feet does it contain?

$$16 \times 10 = 160 \div 12 = 13\tfrac{4}{12}. \quad \textit{Ans.}$$

To find the contents of a plank, joist, or stick of square timber.

RULE. — Multiply the product of the depth and width in inches by the length in feet, and divide the last product by 12; the quotient is the contents in feet, *board measure.*

EXAMPLE. — A joist is 16 feet long, 5 inches wide, and $2\frac{1}{2}$ inches thick; how many feet does it contain, board measure?

$$5 \times 2.5 \times 16 \div 12 = 16\tfrac{8}{12}. \quad \textit{Ans.}$$

To find the solidity of a plank, joist, or stick of square timber.

RULE. — Multiply the product of the depth and width in inches by the length in feet, and divide the last product by 144; the quotient will be the contents in cubic feet.

EXAMPLE. — A stick of timber is 10 by 6 inches, and 14 feet in length; what is its solidity?

$$10 \times 6 = 60 \times 14 = 840 \div 144 = 5\tfrac{5}{6} \text{ feet.} \quad \textit{Ans.}$$

NOTE. — If a board, plank, or joist is narrower at one end than the other, add the two ends together and divide the sum by 2; the quotient will be the mean width. And if a stick of squared timber, whose solidity is required, is narrower at one end than the other $(A + a + \sqrt{Aa}) \div 3 =$ mean area. A and a being the areas of the ends.

To measure round timber.

RULE (IN GENERAL PRACTICE.) — Multiply the length, in feet, by the square of $\frac{1}{4}$ the girt, in inches, taken about $\frac{1}{3}$ the distance from the larger end, and divide the product by 144; the quotient is considered the contents in cubic feet. For a strictly correct rule for measuring round timber, see MENSURATION OF SOLIDS — *Frustum of a Cone.*

EXAMPLE. — A stick of round timber is 40 feet in length, and girts 88 inches; what is its solidity?

$$88 \div 4 = 22 \times 22 = 484 \times 40 = 19360 \div 144 = 134.44 \text{ cub. ft.} \quad \textit{Ans.}$$

The following TABLE *is intended to facilitate the measuring of Round Timber, and is predicated upon the foregoing* RULE.

¼ Girt in Inches.	Area in Feet.	¼ Girt in Inches.	Area in Feet.	¼ Girt in Inches.	Area in Feet.	¼ Girt in Inches.	Area in Feet.
6	.25	12	1.	18	2.25	24	4.
6¼	.272	12¼	1.042	18¼	2.313	24¼	4.084
6½	.294	12½	1.085	18½	2.376	24½	4.168
6¾	.317	12¾	1.129	18¾	2.442	24¾	4.254
7	.34	13	1.174	19	2.506	25	4.34
7¼	.364	13¼	1.219	19¼	2.574	25¼	4.428
7½	.39	13½	1.265	19½	2.64	25½	4.516
7¾	.417	13¾	1.313	19¾	2.709	25¾	4.605
8	.444	14	1.361	20	2.777	26	4.694
8¼	.472	14¼	1.41	20¼	2.898	26¼	4.785
8½	.501	14½	1.46	20½	2.917	26½	4.876
8¾	.531	14¾	1.511	20¾	2.99	26¾	4.969
9	.562	15	1.562	21	3.062	27	5.062
9¼	.594	15¼	1.615	21¼	3.136	27¼	5.158
9½	.626	15½	1.668	21½	3.209	27½	5.252
9¾	659	15¾	1.722	21¾	3.285	27¾	5.348
10	.694	16	1.777	22	3.362	28	5.444
10¼	.73	16¼	1.833	22¼	3.438	28¼	5.542
10½	.766	16½	1.89	22½	3.516	28½	5.64
10¾	.803	16¾	1.948	22¾	3.598	28¾	5.74
11	.84	17	2.006	23	3.673	29	5.84
11¼	.878	17¼	2.066	23¼	3.754	29¼	5.941
11½	.918	17½	2.126	23½	3.835	29½	6.044
11¾	.959	17¾	2.187	23¾	3.917	30	6.25

To find the solidity of a log by help of the preceding TABLE.

RULE. — Multiply the tabular area opposite the corresponding ¼ girt, by the length of the log in feet, and the product will be the solidity in feet.

EXAMPLE. — The ¼ girt of a log is 22 inches, and the length of the log is 40 feet; required the solidity of the log.

$$3.362 \times 40 = 134.48 \text{ cubic feet.} \quad \textit{Ans.}$$

NOTE. — Though custom has established, in a very general way, the preceding method as that whereby to measure round timber, and holds, in most instances, the solidity to be that which the method will give, there seems, if the object sought be the real solidity of the stick, neither accuracy, justice, nor certainty, in the practice.

Thus, in the preceding example, the stick was supposed to be 40 feet in length, and 88 inches in circumference at ¼ the distance from the larger end, and was found, by the method, to contain 134.44 cubic feet: now 88 ÷ 3.1416 = 28 inches, = the diameter at ¼ the distance from the greater base, and retaining this diameter and the length, we may suppose, with sufficient liberality, and without being far from the general run of such sticks, the diameter at the greater base to be 30 inches, and that of the less to be 24 inches, and —

By a correct rule the stick contains —

$30 \times 24 = 720 + 12 = 732 \times .7854 \times 40 = 22996 \div 144 = 159.7$ cubic feet, or 19 per cent. more than given by the method under consideration; and we need hardly add that the nearer the stick approaches to the figure of a cylinder, the wider will be the difference between the truth and the result obtained by the method referred to. Thus, suppose the stick a cylinder, 28 inches in diameter, and 40 feet in length; and we have, by the fallacious rule, as above, 134.44 cubic feet; and —

By a correct method, we have —

$28^2 \times .7854 \times 40 = 24630 \div 144 = 171$ cubic feet, or over 27 per cent. more than furnished by the erroneous mode of practice.

Again: suppose the stick in the form of a cone, 30 inches at the base, and tapering to a point at 150 feet in length; and we have, by a correct rule —

$30^2 \div 3 = 300 \times .7854 \times 150 = 35343 \div 144 = 245.44$ cubic feet; and by the ordinary method of gauging, or the aforementioned practice, we have —

$20 \times 3.1416 = 62.832 \div 4 = 15.708^2 \times 150 = 37011.19 \div 144 = 257$ cubic feet, or nearly 4¾ per cent. more than the stick actually contains.

In short, without taking into account anything for the thickness of the bark, that may be supposed to be on the stick, the method is correct only when the stick tapers at the rate of 5¼ inches diameter per each 10 feet in length, or over ½ inch diameter to each foot in length of the stick.

If, however, we suppose the stick as before, (30 inches at the greater base, 24 inches at the smaller, and 40 feet in length,) and suppose the bark upon it to be 1 inch thick, we shall have, by the usual method, 134.44 cubic feet, as before. And, exclusive of the bark, by a correct method, we shall have.

$\overline{30-2} \times \overline{24-2} = 616 + 12 = 628 \times .7854 \times 40 = 19729 \div 144 = 137$ cubic feet, or only about 2 per cent. more than that furnished us by the usual practice.

The following simple rule for measuring round timber is sufficiently correct for most practical purposes: —

Rule. — Multiply the square of one-fifth of the mean girt, (exclusive of bark,) in inches, by twice the length of the stick in feet, and divide the product by 144; the quotient will be the solidity in feet.

To find the solidity of the greatest rectangular stick that may be cut from a given log, or from a stick of round timber of given dimensions.

Rule. — Multiply the square of the mean diameter of the log, in inches, by half the length of the log, in feet, and divide the product by 144.

Example. — The diameter (exclusive of bark) of the greater base of a stick of round timber is 30 inches, and that of the less base is 24 inches, and the stick is 40 feet in length; required the solidity of the greatest rectangular stick that may be cut from it.

$30 \times 24 + \frac{1}{3}(30-24)^2 = 732 =$ square of mean diameter,* and

$732 \times 20 = 14640 \div 144 = 101\frac{2}{3}$ cubic feet. *Ans.*

* Except in the case of a cylinder, there is a difference betwixt the *mean* diameter of a solid having circular bases, and the *middle* diameter of that solid. The mean diameter reduces the solid to a cylinder; the middle diameter is the diameter midway between the two bases.

NOTE. — The foregoing stick will make —
14640 ÷ 16 = 915 feet of square-edged boards 1 inch thick ;
Or, $101\frac{2}{3} \times 9 = 915$.

To find the solidity of the greatest square stick that may be cut from a given log, or from a stick of round timber of given dimensions.

RULE. — Multiply the square of the diameter of the less end of the log, in inches, by half the length of the log, in feet, and divide the product by 144.

EXAMPLE. — The preceding supposed log will make a square stick containing —

$$24^2 \times \tfrac{40}{2} = 1152 \div 144 = 80 \text{ cubic feet.}$$

Diameter multiplied by .7071 = side of inscribed square.

To find the contents, in Board Measure, of a log, no allowance being made for wane or saw-chip.

RULE. — Multiply the square of the mean diameter, in inches, by the length in feet, and divide the product by 15.28.

Or, Multiply the square of the mean diameter in inches, by the length in feet, and that product by .7854, and divide the last product by 12.

The cubic contents of a log multiplied by 12, equal the contents of the log, board measure.

The convex surface of a Frustum of a Cone = $(C + c) \times \frac{1}{2}$ slant length ; C being the circumference of the greater base, and c the circumference of the less.

GAUGING.

RULES *for finding the capacity in gallons or bushels of different shaped Cisterns, Bins, Casks, &c., and also, by way of examples, for constructing them to given capacities.*

RULE — 1. *When the vessel is rectangular.* Multiply the interior length, breadth, and depth, in feet together, and the product by the capacity of a cubic foot, in gallons or bushels, as desired for its capacity.

RULE — 2. *When the vessel is cylindrical.* Multiply the square of its interior diameter in feet, by its interior depth in feet, and the product by the capacity of a *cylindrical* foot in gallons or bushels, as desired for its capacity.

RULE — 3. *When the vessel is a rhombus or rhomboid.* Multiply its interior length, in feet, its right-angular breath in feet, and its depth in feet together, and the product by the capacity of a cubic foot in the special measure desired for its capacity.

RULE — 4. *When the vessel is a frustum of a cone* — a round vessel larger at one end than the other, whose bases are planes. Multiply the interior diameter of the two ends together, in feet, add ⅓ the square of their difference in feet to the product, multiply the sum by the perpendicular depth of the vessel in feet, and that product by the capacity of a *cylindrical* foot in the unit of measure desired for its capacity.

RULE — 5. *When the vessel is a prismoid or the frustum of any regular pyramid.* To the square root of the product of the areas of its ends in feet, add the areas of its ends in feet, multiply the sum by ⅓ its perpendicular depth in feet, and that product by the capacity of a cubic foot in gallons or bushels, as desired for its capacity.

If it is found more convenient to take the dimensions in inches, do so; proceed as directed for feet, divide the product by 1728, and multiply the quotient by the capacity of the respective foot as directed. Or, multiply the capacity in inches by the capacity of the respective inch in gallons or bushels; — by the quotient obtained by dividing the capacity of the respective foot in gallons or bushels by 1728 — for the contents.

RULE — 6. *When the vessel is a barrel, hogshead, pipe, &c.* Multiply the difference in inches between the bung diameter and head diameter, (interior,) if the staves be

much curved, . .	by .7	*See* page 63.
medium curved, .	by .65	
straighter than medium,	by .6	
nearly straight, .	by .55	

and add the product to the head diameter, taken in inches; then multiply the square of the sum by the length of the cask in inches, and divide the product by the capacity in *cylindrical* inches of a gallon or bushel as desired for the contents. Or, divide the contents in *cylindrical* inches, as above found, by 1728, and multiply the quotient by the capacity of a cylindrical foot in gallons or bushels as desired for its contents. Or, multiply the capacity in cylindrical inches by the capacity of a cylindrical inch, in gallons or bushels, as desired, — that is, by the quotient obtained by dividing the capacity of a cylindrical foot in gallons or bushels, by 1728, for the contents.

The capacity of a

CUBIC FOOT =				CYLINDRICAL FOOT =			
7.4805	Winchester	wine	gallons.	5.8751	Winchester	wine	gallons.
6.1276	Ale		"	4.8126	Ale		"
6.2321	Imperial		"	4.8947	Imperial		"
.80356	Winchester		bushel.	.63111	Winchester		bushel.
.62888	"	heaped	"	.49391	"	heaped	"
.64285	"	1¼ even	"	.50489	"	1¼ even	"
.779	Imperial	"	"	.61183	Imperial		"

EXAMPLE. — Required the capacity in Winchester bushels of a rectangular bin, whose interior length is 12 feet, breadth 6 feet, and depth 5 feet.

$12 \times 6 \times 5 \times .8035 = 289.26$ bushels. *Ans.*

EXAMPLE. — Required the capacity in Winchester wine gallons of a cylindrical can, whose interior diameter is 18 inches, and depth 3 feet.

$18 \times 18 \times 36 \times 5.875 \div 1728 = 39.66$ gallons. *Ans.*

Or, $1.5 \times 1.5 \times 3 \times 5.875 = 39.66$ gallons. *Ans.*

Or, $18 \times 18 \times 36 \times .0034 = 39.66$ gallons. *Ans.*

EXAMPLE. — How many Winchester bushels in 39.66 wine gallons?

$39.66 \times .10742 = 4.26$ bushels. *Ans.*

EXAMPLE. — How many wine gallons in 4.26 Winchester bushels?

$4.26 \times 9.3092 = 39.66$ gallons. *Ans.*

EXAMPLE. — How many wine gallons will a cistern in the form of a frustum of a cone hold, having the interior diameter of one of its ends 6 feet, and that at the other 8 feet, and its perpendicular depth 9 feet?

$8 - 6 = 2$, and $2^2 \div 3 = 1.333 = \frac{1}{3}$ square of dif. of diameters, and
$6 \times 8 + 1.333 = 49.333 \times 9 \times 5.8751 = 2608.55$ gals. *Ans.*

Or, $6 \times 8 + 8^2 + 6^2 = 148 \times \frac{9}{3} \times 5.8751 = 2608.55$ gals. *Ans.*

Or, $(8^3 - 6^3) \div (8 - 6) = 148 \times \frac{9}{3} \times 5.8751 = 2608.55$ gals. *Ans*

Or, $96 - 72 = 24$ and $(24^2 \div 3) = 192$, and
$96 \times 72 + 192 = 7104 \times 108 \times .0034 = 2608.55$ gals. *Ans.*

EXAMPLE. — What is the capacity in Winchester bushels of a cistern whose form is prismoid, the dimensions (interior) of one end being 8 by 6 feet, of the other 4 by 3 feet, and its perpendicular depth 12 feet?

$8 \times 6 = 48 =$ area of one end, and $4 \times 3 = 12 =$ area of the other end; then —

$48 \times 12 = \sqrt{576} = 24 + 48 + 12 = 84 \times \frac{12}{3} \times .80356 = 270$ bushels. *Ans.*

Or, $(8 + 4) \div 2 = 6$, and $(6 + 3) \div 2 = 4.5 =$ mean sectional areas of ends, and

$6 \times 4.5 \times 4 = 4$ area of mean perimeter, then

$8 \times 6 + 4 \times 3 + 6 \times 4.5 \times 4 = 168 \times \frac{12}{6} \times .80356 = 270$ bus. *Ans.*

EXAMPLE. — What must be the depth of a rectangular bin whose length is 12 feet, and breadth 6 feet, to hold 289.26 bushels?

$$289.26 \div (12 \times 6 \times .80356) = 5 \text{ feet.} \quad \textit{Ans.}$$

EXAMPLE. — A cylindrical can, whose depth is to be 36 inches, is required to be made that will hold 40 gallons; what must be the diameter of the can?

$$40 \div (3 \times 5.8751) = \sqrt{2.27} = 1.506 \text{ feet.} \quad \textit{Ans.}$$
$$\text{Or, } 40 \div (36 \times .0034) = \sqrt{326.8} = 18.07 \text{ inches.} \quad \textit{Ans.}$$

EXAMPLE. — A cylindrical can, whose interior diameter is to be 18 inches, is required that will hold 40 gallons; what must be the interior depth of the can?

$$40 \div (18^2 \times .0034) = 36.31 \text{ inches.} \quad \textit{Ans.}$$
$$\text{Or, } 40 \div (1.5^2 \times 5.8751) = 3.026 \text{ feet.} \quad \textit{Ans.}$$

EXAMPLE. — A cistern is to be built in the form of a frustum of a cone, that will hold 1800 gallons, and the diameter of one of its ends is to be 5 feet, and that of the other $7\frac{1}{2}$ feet; what must be the depth?

$7.5 - 5 = 2.5$, and $2.5^2 \div 3 = 2.0833 = \frac{1}{3}$ square of difference of diameter, and

$$1800 \div (7.5 \times 5 + 2.0833) \times 5.8751 = 7.74 \text{ feet.} \quad \textit{Ans.}$$

$$\text{Or, } 1800 \div \left(\frac{7.5 \times 5 + 7.5^2 + 5^2}{3} \times 5.8751\right) = 7.74 \text{ feet.} \quad \textit{Ans.}$$

EXAMPLE. — The form, capacity, depth, and diameter of one end being determined on, and being as above, what must be the diameter of the other end?

$\frac{c}{\frac{1}{3}h} - \frac{3}{4}d^2 = y$, c being the solidity in cylindrical measurement, h

the depth, d the diameter of the given end or base, and y a quantity the square root of which is the sum of the required base and half the given base; then

$$1800 \div 5.8751 = 306.378 = \text{solidity in cylindrical feet, and}$$

$$306.378 \div \tfrac{7.74}{3} = 118.75 - (5^2 \div \tfrac{4}{3}) = \sqrt{100} = 10 - \tfrac{5}{2} = 7.5$$ feet. *Ans.*

Example. — A measure is to be built in the form of a frustum of a cone, that will hold exactly 1 wine gallon, and the diameter of one of its ends is to be 4 inches, and that of the other 6 inches; what must be its depth?

$$1 \div (6 \times 4 + 1\tfrac{1}{3}) \times .0034 = 11.61 \text{ inches. } \textit{Ans.}$$

$$\text{Or, } \frac{231}{.7854} \div \frac{6 \times 4 + 6^2 + 4^2}{3} = 11.61 \text{ inches. } \textit{Ans.}$$

Example. — A measure in the form of a frustum of a cone holds 1 wine gallon; the diameter of one of its ends is 6 inches, and its depth is 11.61 inches; what is the diameter of the other end?

$$\tfrac{231}{.7854} = 294.1176 \div \tfrac{11.61}{3} = 76 - (6^2 \div \tfrac{4}{3}) = \sqrt{49} = 7 - \tfrac{6}{2}$$ $= 4$ inches. *Ans.*

CASK GAUGING.

Cask-gauging, in a general sense, is a practical art, rather than a scientific achievement or problem, and makes no pretensions to strict accuracy with regard to the conclusions arrived at. The aim is, by means of a few satisfactory measurements taken of the outside, and an estimate of the probable mean thickness of the material of which the cask is composed (of which there must always remain some doubt), or by means of a few measurements taken of the inside, to determine, 1st, the capacity of the cask, and, 2d, the ullage, or capacity of the occupied or unoccupied space in a cask but partly full. And the Rule (Rule 6, page 60), which reduces the supposed cask, or cask of supposed curvature, to a cylinder, is as practically correct for the capacity of ordinary casks, as any rule, or set of rules, that can be offered for general purposes.

Casks have no fixed form of their own, to which they severally and collectively correspond, nor are they in any considerable degree in conformity with any regular geometrical figure.

Some casks — a few — those having their staves much curved throughout their entire length, are nearest in keeping with the *middle frustum of a spheroid;* others, slightly less curved than the preceding, correspond in a considerable degree to the *middle frustum of a parabolic spindle;* others, again — those having very little longitudinal curvature of stave to their semi-lengths — are nearly in keeping with the *equal frustums of a paraboloid;* and others — a very few — those whose staves are straight from the bung diameter to the heads, or equal to that form, are in accordance with the *equal frustums of a cone.*

The *gauging rod*, which is intended to be correct for casks of the most common form, gives for all casks, as may be seen in one of the following EXAMPLES, a solidity slightly greater (about $2\frac{1}{2}$ per cent.) than would be obtained by supposing the cask in conformity with the third figure above alluded to.

The RULE for finding the contents of a cask, *by four dimensions*, hereafter to be given, is intended as a general Rule for all casks, and, when the diameter midway between the bung and head can be accurately ascertained, will lead to a very close approach to the truth.

From the length of a cask, taken from outside to outside of the heads, with callipers, it is usual to deduct from 1 to 2 inches, to correspond with the thickness of the heads, according to the size of the cask, and the remainder is taken as the length of the interior.

To the diameter of each head, taken externally, from $\frac{1}{4}$ inch to $\frac{3}{10}$ inch should be added for common-sized barrels, $\frac{4}{10}$ inch for 40 gallon casks, and from $\frac{1}{2}$ inch to $\frac{6}{10}$ inch for larger casks, to correspond with the interior diameters of the heads.

If the staves are of uniform thickness, any sectional diameter of a cask may be nearly or quite ascertained, by dividing the circumference at that place by 3.1416, and subtracting twice the thickness of the stave from the quotient.

For obtaining the diagonal of a cask by mathematical process, — the interior length, &c. &c. — see *Rules*, below.

In the following formulas D denotes the bung diameter, d the head diameter, and l the length of the cask.

The solidity of any cask is equal to its length multiplied by the square of its mean diameter multiplied by ·7854.

To calculate the contents of a cask from four dimensions.

RULE. — To the square of the bung diameter add the square of the head diameter, and the square of double the diameter midway between the bung and head, and multiply the sum by $\frac{1}{6}$ the length of the cask, for its *cylindrical* contents; the product multiplied by .0034 expresses the contents in wine gallons.

EXAMPLE. — The length of the cask is 40 inches, its bung diameter 28 inches, head diameter 20 inches, and the diameter midway be tween the bung and head is 25.6 inches; how many gallons' capacity has the cask?

$20^2 + 28^2 + \overline{25.6 \times 2}^2 = 3805.44 \times \frac{40}{6} \times .0034 = 86.26$ gals. *Ans.*

$$(D^2 + d^2 + \overline{2\,m}^2) \times \tfrac{1}{6}\,l \times .7854 = \text{cubic contents.}$$

$$\frac{D^2 + d^2 + \overline{2\,m}^2}{6} = \text{square of mean diameter.}$$

By RULE 6, p. 68, this cask will hold —

$28 - 20 = 8 \times .65 = 5.2 + 20 = 25.2 \times 25.2 \times 40 \times .0034 = 86.36$ gallons.

When the cask is in the form of the middle frustum of a spheroid

$$\tfrac{2}{3}\,D^2 + \tfrac{1}{3}\,d^2 = \text{square of mean diameter.}$$

And a cask of this form, having the same head diameter, bung diameter, and length as the preceding, will hold —

$$\frac{2 \times 28^2 + 20^2}{3} \times 40 \times .0034 = 89.216 \text{ gallons.}$$

When the cask is in the form of the middle frustum of a parabolic spindle.

$$\overline{\tfrac{8}{15}\,D^2 + \tfrac{1}{3}\,d^2} - \tfrac{2}{15}\,(D \backsim d)^2 = \text{square of mean diameter.}$$

And a cask of this form, having the same head diameter, bung diameter, and length as the preceding, will hold —

$522\tfrac{2}{3} + 133\tfrac{1}{3} = 656 - 8.533 = 647.467 \times 40 \times .0034 = 88.055$ gals.

When the cask is in the form of two equal frustums of a paraboloid.

$$\tfrac{1}{2}\,D^2 + \tfrac{1}{2}\,d^2 = \text{square of mean diameter.}$$

And a cask of this form, having the same head diameter, bung diameter, and length as the preceding, will hold —

$$\frac{28^2 + 20^2}{2} \times 40 \times .0034 = 80.51 \text{ gallons.}$$

When the cask is in the form of the equal frustums of a cone.

$$\overline{\tfrac{1}{2}\,D^2 + \tfrac{1}{2}\,d^2} - \tfrac{1}{6}\,(D \backsim d)^2 = \text{square of mean diameter.}$$

Or, $\tfrac{1}{3}D^2 + \tfrac{1}{3}d^2 + \tfrac{1}{3}Dd =$ " " " "

Or, $D \times d + \tfrac{1}{3}(D \backsim d)^2 =$ " " " "

And a cask of this form, having the same head diameter, bung diameter, and length as the preceding, will hold —

$\overline{28 \times 20 + 21\tfrac{1}{3}} \times 40 \times .0034 = 79.06$ gals

To find the contents of a cask the same as would be given by the gauging rod.

The *gauging rod* is constructed upon the principle that the cube of the diagonal of a cask, in inches, multiplied by $\frac{144}{64000}$, equals the contents of the cask, in Imperial gallons.

The contents in wine gallons of either of the aforementioned casks, therefore, by the gauging rod, would be —

$$\overline{31.241}^3 \times .0027 = 82\tfrac{1}{3} \text{ gals.}$$

The decimal coëfficient to take the place of .0027, for finding the contents of a cask in the form of the middle frustum of a spheroid = .002926; and for finding the contents of a cask in the form of the equal frustums of a cone = .002593. And between these extremes lies the decimal for other casks, or casks of intervening figures.

To find the diagonal of a cask, when the interior is inaccessible.

Rule. — From the bung diameter subtract half the difference of the bung and head diameters, and to the square of the remainder add the square of half the length of the cask, and the square root of the sum will be the diagonal.

Example. — What is the diagonal of a cask whose bung diameter is 28 inches, head diameter 20 inches, and length 40 inches?

$$28 - 20 = 8 \div 2 = 4, \text{ and } 28 - 4 = 24, \text{ then}$$

$$\sqrt{(24^2 + 20^2)} = 31.241 \text{ inches.} \quad \textit{Ans.}$$

To find the length of a cask, the head diameter, bung diameter and diagonal being given.

$$\sqrt{\left(\text{diagonal}^2 - \overline{D - \frac{D \backsim d}{2}}^2\right)} = \tfrac{1}{2}\, l.$$

And the interior length of a cask, whose interior head diameter, bung diameter and diagonal, are as the preceding, will be

$$\sqrt{(31.241^2 - 24^2)} = 20 \times 2 = 40 \text{ inches.}$$

To find the solidity of a sphere.

$$D^2 \times \tfrac{2}{3}\, D \times .7854 = \text{cubic contents, D being the diameter.}$$

To find the solidity of a spherical frustum.

$$\left(\tfrac{2}{3}\, h^2 + \frac{b^2 + d^2}{2}\right) \times h \times .7854 = \text{cubic contents, } b \text{ and } d \text{ being the bases, and } h \text{ the height.}$$

Note. — For Rules in detail pertaining to the foregoing figures, and for other figures, see Mensuration of Solids.

ULLAGE.

The *ullage* or *wantage* of a cask is the quantity the cask lacks of being full.

To find the ullage of a standing cask, when the cask is half full or more.

RULE. — To the square of the head diameter, add the square of the diameter at the surface of the liquor, and the square of twice the diameter midway between the surface of the liquor and the upper head, and divide the sum by 6; the quotient, multiplied by the distance from the surface of the liquor to the upper head, multiplied by .0034, will give the ullage in wine gallons.

EXAMPLE. — The diameters are as follows — at the upper head, 20 inches; at the surface of the liquor, 22 inches; and at a point midway between these, 21¼ inches; and the distance from the upper head to the surface of the liquor is 5 inches; required the ullage.

$(20^2 + 22^2 + \overline{21.25 \times 2}^2) \div 6 = 448.37 \times 5 \times .0034 = 7.62$ gallons. *Ans.*

When the cask is standing, and less than half full, to find the ullage.

RULE. — Make use of the bung diameter in place of the head diameter, and proceed in all respects as directed in the last *Rule*, and add the quantity found to half the capacity of the cask; the sum will be the ullage.

EXAMPLE. — The bung diameter is 28 inches; the diameter at the surface of the liquor, below the bung, is 26 inches; the diameter midway between the bung and the surface of the liquor is 27.3 inches; and the distance from the surface of the liquor to the bung diameter is 5 inches; required the quantity the cask lacks of being half full; and also the ullage of the cask, its capacity being 86.26 gallons.

$(28^2 + 26^2 + \overline{27.3 \times 2}^2) \div 6 = 740.2 \times 5 \times .0034 = 12.58$ gallons less than ½ full. *Ans.*

And, $86.26 \div 2 = 43.13 + 12.58 = 55.73$ gallons ullage. *Ans.*

When the cask is upon its bilge, and half full or more, to find the ullage.

RULE. — Divide the distance from the bung to the surface of the liquor — (the height of the empty segment) — by the whole bung diameter, and take the quotient as the height of the segment of a circle whose diameter is 1, and find the area of the segment; multiply the area by the capacity of the cask, in gallons, and that product by 1.25; the last product will be the ullage, in gallons, as found by the aid of the *wantage-rod;* and will be correct for casks of the most common form.

NOTE. — The area of the segment of a circle =

(ch'd $\frac{1}{2}$ arc + $\frac{1}{3}$ ch'd $\frac{1}{2}$ arc + ch'd seg.) × height seg. × $\frac{4}{10}$*, very nearly.

And, having the diameter of the circle and the height of the segment given, the chord of half the arc, and the chord of the segment may be found, thus —

radius — height = cosine ; radius^2 — cosine^2 = sine^2; $\sqrt{}$ *(sine^2)* × 2 = *ch'd of seg.*

sine^2 + height seg.^{2} = ch'd $\frac{1}{2}$ arc^2, and $\sqrt{}$ *(ch'd $\frac{1}{2}$ arc^2) = ch'd $\frac{1}{2}$ arc.*

EXAMPLE. — The bung diameter is 28 inches, the height of the empty segment 5.6 inches, and the capacity of the cask 86.26 gallons; required the ullage of the cask, in gallons.

$5.6 \div 28 = .2 =$ height of seg., diameter as 1.

$1 \div 2 = .5 =$ radius.

$.5 - .2 = .3 =$ cosine.

$.5^2 - .3^2 = .16 =$ sine^2, or square of half the base of the segment.

$\sqrt{.16} = .4 \times 2 = .8 =$ chord of segment, or base of segment.

$.4^2 + .2^2 = .2 =$ square of chord of half the arc.

$\sqrt{.2} = .4472 =$ chord of half the arc, then —

$.4472 \div 3 = .1491$, and $\overline{.1491 + .4472 + .8} \times .2 \times \frac{4}{10} = .1117$, area of segment, and

$.1117 \times 86.26 \times 1.25 = 12$ gallons. *Ans.*

When the cask is upon its bilge, and less than half full, to find the ullage.

RULE. — Divide the depth of the liquor by the bung diameter, and proceed in all respects as directed in the last Rule; then subtract the quantity found from the capacity of the cask, and the difference will be the ullage of the cask.

To find the quantity of liquor in a cask by its weight.

EXAMPLE. — The weight of a cask of proof spirits is 300 lbs., and the weight of the empty cask (*tare*) is 32 lbs. How many gallons are there of the liquor?

$300 - 32 = 268 \div 7.732 = 34\frac{2}{3}$ gallons. *Ans.*

Customary Rule by Freighting Merchants, for finding the cubic measurement of casks.

Bung diameter^2 × $\frac{4}{5}$ length of cask = cubic measurement.

NOTE. — One cubic foot contains 7.4805 wine gallons.

* For several Rules in detail, for finding the area of the segment of a circle, see GEOMETRY — *Mensuration of Superficies.*

TONNAGE.

GOVERNMENT MEASUREMENT.*

$$\frac{\text{length} - \frac{3}{5}\text{ breadth} \times \text{breadth} \times \text{depth}}{95} = \text{tonnage}$$

In a double-decked vessel, the length is reckoned from the fore part of the main stem to the after side of the sternpost above the upper deck; the breadth is taken at the broadest part above the main wales, and half this breadth is taken for the depth.

In a single-decked vessel the length and breadth are taken as for a double-decked vessel, and the distance between the ceiling of the hold and the under side of the deck plank is taken as the depth.

Example. — The length of a double-decked vessel is 260 feet, and the breadth is 60 feet; required the tonnage.

$260 - \frac{60 \times 3}{5} = 224 \times 60 \times \frac{60}{2} = 403200 \div 95 = 4244.2$ tons. *Ans.*

Example. — The length of a single-decked vessel is 180 feet, the breadth 34 feet, and depth 18 feet; required the tonnage.

$180 - \frac{3}{5}$ of $34 = 159.6 \times 34 \times 18 \div 95 = 1028.16$ tons. *Ans.*

CARPENTER'S MEASUREMENT.

For a double-decked —

$$\frac{\text{length of keel} \times \text{breadth main beam} \times \frac{1}{2}\text{ breadth}}{95} = \text{tonnage.}$$

For a single-decked —

$$\frac{\text{length of keel} \times \text{breadth main beam} \times \text{depth of hold}}{95} = \text{tonnage}$$

* The set of rules legalized by the Congress of 1865 for ascertaining the nominal or Government tonnage of vessels are not inserted in this work, partly because of their great lengths in detail, and the multiplicity of the mechanical admeasurements required, and partly because they can be of use to but a few individuals; and to those they are furnished by the Government.

OF CONDUITS OR PIPES.

Pressure of Water in Vertical Pipes, &c.

h = height of column in inches ; o = circumference of column in inches; t = thickness of pipe in inches equal in strength to lateral pressure at base of column ; w = weight of a cubic inch of water in pounds ; C = cohesive strength in pounds per inch area of transverse section of the material of which the pipe is composed — TABLE, p. 74.

ho = area of interior of pipe in inches ; hw = pressure in pounds per square inch at the base of the column, or maximum lateral pressure in pounds per square inch on the pipe tending to burst it ; how = maximum lateral pressure in pounds on the pipe, tending to burst it at the bottom ; and $how \div 2$ = mean lateral pressure in pounds on the pipe, or pressure in pounds on the pipe tending to burst it at half the height of the column.

$$how \div C = t; \quad how \div t = C; \quad Ct \div ow = h; \quad Ct \div hw = o.$$

NOTE. — The *reliable* cohesion of a material is not above $\frac{1}{3}$ its ultimate force, as given in the Table of Cohesive Forces. By experiment, it has been found that a cast iron pipe 15 inches in diameter and $\frac{3}{4}$ of an inch thick, will support a head of water of 600 feet ; and that one of the same diameter made of oak, and two inches thick, will support a head of 180 feet : 12000 lbs. per square inch for cast iron, 1200 for oak, 750 for lead, are counted safe estimates. The ultimate cohesion of an alloy, composed of lead 8 parts and zinc 1 part, is 3000 pounds per square inch.

Concerning the Discharge of Pipes, &c.

Small pipes, whether vertical, horizontal, or inclined, under equal heads, discharge proportionally less water than large ones. That form of pipe, therefore, which presents the least perimeter to its area, other things being equal, will give the greatest discharge. A round pipe, consequently, will discharge more water in a given time than a pipe of any other form, of equal area.

The greater the length of a pipe discharging vertically, the greater the discharge. Because the friction of the particles against its sides, and consequent retardation, is more than overcome by the gravity of the fluid.

The greater the length of a pipe discharging horizontally, the less proportionally will be the discharge. The proportion compared with a less length is in the inverse ratio of the square root of the two lengths, nearly.

Other things being equal, rectilinear pipes give a greater discharge than curvilinear, and curvilinear greater than angular. The head, the diameters and the lengths being the same, the time occupied in passing an equal quantity of water through a straight pipe is 9, through one curved to a semicircle 10, and through one having one right angle, otherwise straight, 14. All interior inequalities and roughness should be avoided.

It has been ascertained that a velocity of 60 feet a minute (1 foot a second) through a horizontal pipe, 4 inches in diameter and 100 feet in length, is produced by a head $2\frac{1}{4}$ inches, only $\frac{1}{4}$ of an inch above the upper surface of the orifice; and that, to maintain an equal velocity through a pipe similarly situated, of equal length, having a diameter of $\frac{1}{2}$ inch only, a head of $1\frac{5}{12}$ feet is required. To increase the velocity through the last mentioned pipe to 2 feet a second, requires a head $4\frac{10}{12}$ feet; to 3 feet, a head of $10\frac{1}{12}$; to 4 feet, a head of $17\frac{10}{12}$, &c.

From the foregoing, the following, it is believed, reliable rules, are deduced.

To find the velocity of water passing through a straight horizontal pipe of any length and diameter, the head, or height of the fluid above the centre of the orifice, being known.

Rule. — Multiply the head, in feet, by 2500, and divide the product by the length of the pipe, in feet, multiplied by 13.9, divided by the interior diameter of the pipe in inches; the square root of the quotient will be the velocity in feet per second.

Example. — The head is 6 feet, the length of the pipe 1340 feet, and its diameter 5 inches; required the velocity of the water passing through it.

$2500 \times 6 = 15000 \div (\frac{1340 \times 13.9}{5}) = \sqrt{4.03} - = 2$ feet per second. *Ans.*

To find the head necessary to produce a required velocity through a pipe of given length and diameter.

Rule. — Multiply the square of the required velocity, in feet, per second, by the length of the pipe multiplied by the quotient obtained by dividing 13.9 by the diameter of the pipe in inches, and divide the product thus obtained by 2500; the quotient will be the head in feet.

Example. — The length of a pipe lying horizontal and straight is 1340 feet, and its diameter is 5 inches; what head is necessary to cause the water to flow through it at the rate of 2 feet a second?

$2^2 \times 1340 \times \frac{13.9}{5} \div 2500 = 6$ feet. *Ans.*

To find the quantity of water flowing through a pipe of any length and diameter.

Rule. — Multiply the velocity in feet per second by the area of the discharging orifice, in feet, and the product is the quantity in cubic feet discharged per second.

Example. — The velocity is 2 feet a second, and the diameter of the pipe 5 inches; what quantity of water is discharged in each second of time?

$5 \div 12 = .4166$, and $.4166^2 \times .7854 \times 2 = .273$ cubic foot. *Ans.*

MISCELLANEOUS PROBLEMS.

To find the specific gravity of a body heavier than water.

Rule. — Weigh the body in water and out of water, and divide the weight out of water by the difference of the two weights.

Example. — A piece of metal weighs 10 lbs. in atmosphere, and but 8¼ in water; required its specific gravity.

$10 - 8.25 = 1.75$, and $10 \div 1.75 = 5.714$. *Ans.*

To find the specific gravity of a body lighter than water.

Rule. — Weigh the body in air; then connect it with a piece of metal whose weight, both in and out of water, is known, and of sufficient weight that the two will sink in water; and find their combined weight in water; then divide the weight of the body in air by the weight of the two substances in air, less the sum of the difference of the weight of the metal in air and water and the combined weight of the two substances in water, and the quotient will be the specific gravity sought.

Example. — The combined weight, in water, of a piece of wood, and piece of metal, is 4 lbs.; the wood weighs in atmosphere 10 lbs.; and the metal in atmosphere 12, and in water 11 lbs.; required the specific gravity of the wood.

$10 \div (10 + 12 - \overline{12 \backsim 11 + 4}) = .588$. *Ans.*

To find the specific gravity of a fluid.

Rule. — Multiply the known specific gravity of a body by the difference of its weight in and out of the fluid, and divide the product by its weight out of the fluid; the quotient will be the specific gravity of the fluid in which the body is weighed.

Example. — The specific gravity of a brass ball is 8.6; its weight in atmosphere is 8 oz., and in a certain fluid 7¼ oz.; required the specific gravity of the fluid.

$8 - 7.25 = .75$, and $8.6 \times .75 = 6.45$, and $6.45 \div 8 = .806$. *Ans.*

To find the proportion of one to the other of two simples forming a compound, or the extent to which a metal is debased, (the metal and the alloy used being known.)

The Rule strictly bears upon that of *Alligation Alternate*, which see.

Example. — The specific gravity of gold is 19.258, and that of copper, 8.788; an article composed of the two metals, has a specific gravity of 18; in what proportion are the metals mixed?

$18 \backsim 19.258 \times 8.788 = 11.055$

$18 \backsim 8.788 \times 19.258 = 177.4$, then

$\overline{11.055 + 177.4} : 11.055 :: 18 = 1.056$ copper, $\}$ *Ans.*
$\overline{11.055 + 177.4} : 177.4 :: 18 = 16.944$ gold. $\}$
Or, $18 - 1.056 = 16.944$ gold. Copper to gold as 1 to 16.04 +

To find the lifting power of a balloon.

RULE. —Multiply the capacity of the balloon, in feet, by the difference of weight between a cubic foot of atmosphere and a cubic foot of the gas used to inflate the balloon, and the product is the weight the balloon will raise.

EXAMPLE. —A balloon, whose diameter is 24 feet, is inflated with hydrogen; what weight will it raise?

Specific gravity of air is 1, weight of a cubic foot 527.04 grains; specific gravity of hydrogen is .0689.
$527.04 \times .0689 = 36.31$ grains = weight of 1 cubic foot of hydrogen.
$527.04 - 36.31 = 490.73$ grs. = dif. of weight of air and hydrogen.
$24^3 \times .5236 = 7238.24$ = capacity in cubic feet of balloon.
Then, $7238.24 \times 490.73 = 3552021$ grs. $= \frac{3552021}{7000} = 507\frac{4}{10}$ lbs.
Ans.

To find the diameter of a balloon that shall be equal to the raising of a given weight.

The weight to be raised is $507\frac{4}{10}$ lbs.

$\overline{507.4 \times 7000} \div 490.73 = 7238.24$, and $7238.24 \div .5236 = \sqrt[3]{13824}$
$= 24$ feet. *Ans.*

To find the thickness of a concave or hollow metallic ball or globe, that shall have a given buoyancy in a given liquid.

EXAMPLE. —A concave globe is to be made of brass, specific gravity 8.6, and its diameter is to be 12 inches; what must be its thickness that it may sink exactly to its centre in pure water?

Weight of a cubic inch of water .036169 lb.; of the brass .3112 lb.
Then, $\overline{12^3 \times .5236 \times .036169} \div 2 = 16.3625$ cubic inches of water to be displaced.
$16.3625 \div .3112 = 52.5787$ cubic inches of metal in the ball.
$\overline{12^2 \times 3.1416} = 452.39$ square inches of surface of the ball.
And, $52.5787 \div 452.39 = .1162 + = \frac{1}{9}$ inch thick, full. *Ans.*

To cut a square sheet of copper, tin, etc., so as to form a vessel of the greatest cubical capacity the sheet admits of.

RULE. —From each corner of the sheet, at right angles to the side, cut $\frac{1}{6}$ part of the length of the side, and turn up the sides till the corners meet.

Comparative Cohesive Force of Metals, Woods, and other substances, Wrought Iron (medium quality) being the unit of comparison, or 1; *the cohesive force of which is* 60000 *lbs. per inch. transverse area.*

Wrought iron,	1.00	Ash, white,	.23
" " wire,	1.71	" red,	.30
Copper, cast,	.40	Beech,	.19
" wire,	.76	Birch,	.25
Gold, cast,	.34	Box,	.33
" wire,	.51	Cedar,	.19
Iron, cast, (average).	.38	Chestnut, sweet,	.17
Lead, "	.015	Cypress,	.10
" milled,	.055	Elm,	.22
Platinum, wire,	.88	Locust,	.34
Silver, cast,	.66	Mahogany, best,	.36
" wire,	.68	Maple,	.18
Steel, soft,	2.00	Oak, Amer., white,	.19
" fine,	2.25	Pine, pitch,	.20
Tin, cast block,	.083	Sycamore,	.22
Zinc, "	.043	Walnut,	.30
" sheet,	.27	Willow,	.22
Brass, cast,	.75	Ivory,	.27
Gun metal,	.50	Whalebone,	.13
Gold 5, copper 1,	.83	Marble,	.15
Silver 5, " 1,	.80	Glass, plate,	.16
Brick,	.05	Hemp fibres, glued,	1.53
Slate,	.20		

The *strength* of white oak to cast iron, is as 2 to 9.
The *stiffness* of " " " " is as 1 to 13.

To determine the weight, or force, in pounds, necessary to tear asunder a bar, rod, or piece of any of the above named substances, of any given transverse area:

Rule. — Multiply the comparative cohesive force of the substance, as given in the table, by the cohesive force per square inch, area of cross section (60000 lbs.) of wrought iron, which gives the cohesive force of 1 square inch area of cross section of the substance whose power is sought to be ascertained, and the product of 1 square inch thus found, multiplied by area of cross section, in inches, of the rod, piece, or bar itself, gives the cohesive force thereof.

Alloys having a tenacity greater than the sum of their constituents.

Swedish copper 6 pts., Malacca tin 1; tenacity per sq. inch, 64000 lbs
Chili copper 6 pts., Malacca tin 1; " " " 60000 "
Japan copper 5 pts., Banca tin 1; " " " 57000 "
Anglesea copper 6 pts., Cornish tin 1; " " " 41000 "

Common block-tin 4 pts., lead 1, zinc 1; tenacity per sq. in., 13000 lbs.
Malacca tin 4 pts., regulus of antimony 1; " " " 12000 "
Block-tin 3 pts., lead 1 part; " " " 10000 "
Block-tin 8 pts., zinc 1 part; " " " 10200 "
Zinc 1 part, lead 1 part; " " " 4500 "

Alloys having a density greater than the mean of their constituents.

GOLD with *antimony, bismuth, cobalt, tin*, or *zinc*.
SILVER with *antimony, bismuth, lead, tin*, or *zinc*.
COPPER with *bismuth, palladium, tin*, or *zinc*.
LEAD with *antimony*.
PLATINUM with *molybdinum*.
PALLADIUM with *bismuth*.

Alloys having a density less than the mean of their constituents.

GOLD with *copper, iron, iridium, lead, nickel*, or *silver*.
SILVER with *copper* or *cobalt*.
IRON with *antimony, bismuth*, or *lead*.
TIN with *antimony, lead*, or *palladium*.
NICKEL with *arsenic*.
ZINC with *antimony*.

RELATIVE POWER OF DIFFERENT METALS TO CONDUCT ELECTRICITY,

(*the mass of each being equal.*)

Copper,	1000	Platinum,	188
Gold,	936	Iron,	158
Silver,	736	Tin,	155
Zinc,	285	Lead,	83

LINEAR DILATION OF SOLIDS BY HEAT.

Length which a bar heated to 212° has greater than when at the temperature of 32°.

Brass, cast,	.0018671	Iron, wrought,	.0012575
Copper,	.0017674	Lead,	.0028568
Fire brick,	.0004928	Marble,	.0011016
Glass,	.0008545	Platinum,	.0009342
Gold,	.0014880	Silver,	.0020205
Granite,	.0007894	Steel,	.0011898
Iron, cast,	.0011111	Zinc,	.0029420

NOTE. — To find the *surface* dilation of any particular article, double its linear dilation, and to find the dilation in *volume*, triple it. To find the elongation in linear inches per linear foot, of any particular article, multiply its respective linear dilation, as given in the TABLE, by 12.

MELTING POINT OF METALS AND OTHER BODIES.

Lime, *palladium*, *platinum*, *porcelain*, *rhodium*, *silex*, may be melted by means of strong lenses, or by the hydro-oxygen blowpipe. *Cobalt*, *manganese*, *plaster of Paris*, *pottery*, *iron*, *nickel*, &c., at from 2300° to 3250° Fahrenheit; others as follows: —

	Degrees Fah.		Degrees Fah.
Antimony,	809	Nitre,	660
Beeswax, bleached,	155	Silver,	1873
Bismuth,	506	Solder, common,	475
Brass,	1900	" plumbers',	360
Copper,	1996	Sugar,	400
Glass, flint,	1178	Sulphur,	226
Gold,	2216	Tallow,	127
Lead,	612	Tin,	442
Mercury,	—39	Zinc,	680

Cast iron thoroughly melts at	2386
Greatest heat of a smith's forge, (com.)	2346
Welding heat of iron,	1892
Iron red hot in twilight,	884
Lead 1, tin 1, bismuth 4, melts at	201
Lead 2, tin 3, bismuth 5, " "	212

RELATIVE POWER OF DIFFERENT BODIES TO RADIATE HEAT.

Water,	100	Lead, bright,	19
Copper,	12	Mercury,	20
Glass,	90	Paper, white,	100
Ice,	85	Silver,	12
India ink,	88	Tin, blackened,	100
Iron, polished,	15	" clean,	12
Lampblack,	100	" scraped,	16

NOTE. — The power of a body to *reflect* heat is inverse to its power of radiation.

BOILING POINT OF LIQUIDS.

Barometer at 30 in.

Acid, nitric,	253°	Oils, essential, avg.,	318°
" sulphuric,	600°	" turpentine,	316°
Alcohol, *anhyd.*,	168.5°	" linseed,	640°
" 90 per cent.,	174°	Phosphorus,	554°
Ether, sulph.,	97°	Sulphur,	560°
Mercury,	656°	Water,	212°

NOTE. — Barometer at 31 inches, water boils at 213°.57; at 29, it boils at 210°.38; at 28, it boils at 208°.69; at 27, it boils at 206°.85, and in *vacuo* it boils at 88°. No liquid, under pressure of the atmosphere alone, can be heated above its boiling point. At that point the steam emitted sustains the weight of the atmosphere.

FREEZING POINT OF LIQUIDS.

Acid, nitric,	—55°	Oil, linseed, avg.,	—11°
" sulphuric,	1°	Proof spirits,	—7°
Ether,	—47°	Spirits turpentine,	16°
Mercury,	—39°	Vinegar,	28°
Milk,	30°	Water,	32°
Oil, cinnamon,	30°	Wine, strong,	20°
" fennel,	14°	Rapeseed Oil,	25°
" olive,	36°		

NOTE. — Water expands in freezing .11, or $\frac{1}{9}$ its bulk.

EXPANSION OF FLUIDS BY BEING HEATED FROM 32° TO 212°, F.

Atmospheric air, $\frac{1}{480}$ per each degree, =	.375
Gases, all kinds, $\frac{1}{480}$ " " "	
Mercury, exposed,	.018
Muriatic acid, (sp. gr. 1.137,)	.060
Nitric acid, (sp. gr. 1.40,)	.110
Sulphuric acid, (sp. gr. 1.85,)	.060
" ether, — to its boiling point,	.070
Alcohol, (90 per cent.,) " "	.110
Oils, fixed,	.080
" turpentine,	.070
Water,	.046

RELATIVE POWER OF SUBSTANCES TO CONDUCT HEAT.

Gold,	1000	Zinc,	363
Silver,	973	Tin,	304
Copper,	898	Lead,	180
Platinum,	381	Porcelain,	12
Iron,	374	Fire brick,	11

NOTE. — Different woods have a conducting power in ratio to each other, as is their respective specific gravities, the more dense having the greater.

METALS IN ORDER OF DUCTILITY AND MALLEABILITY.

Ductility.	*Malleability.*
1. Platinum.	1. Gold.
2. Gold.	2. Silver.
3. Silver.	3. Copper.
4. Iron.	4. Tin.
5. Copper.	5. Platinum.
6. Zinc.	6. Lead.
7. Tin.	7. Zinc.
8. Lead.	8. Iron.

Quantity per cent. by weight of Nutritious Matter contained in different articles of Food.

Articles.	per ct.	Articles.	per ct.
Lentils, ...	94	Oats, ...	74
Peas, ...	93	Meats, avg., ...	35
Beans, ...	92	Potatoes, ...	25
Corn, (maize,)	89	Beets, ...	14
Wheat, ...	85	Carrots, ...	10
Barley, ...	83	Cabbage, ...	7
Rice, ...	88	Greens, ...	6
Rye, ...	79	Turnips, white, . . .	4

Specific gravity, and quantity per cent., by volume, of Absolute Alcohol contained, necessary to constitute the following named unadulterated articles.

Articles.	Sp. grav. 60°, b. 30.	Per cent. of Alcohol.
Absolute Alcohol, (*anhydrous*,)	.7939	100
Alcohol, highest by distillation, ...	.825	92.6
" commercial standard, ...	.8335	90
Proof Spirits, — standard, ...	.9254	54

Quantity per cent., by volume, (general average) of Absolute Alcohol contained in different pure or unadulterated Liquors, Wines, &c.

Liquors, &c.	per cent.	Wines.	per cent.
Rum, ...	50	Port, . . .	22
Brandy, ...	50	Madeira, . . .	20
Gin, Holland, ...	48	Sherry, ...	18
Whiskey, Scotch, . . .	50	Lisbon, ...	17
" Irish, . . .	50	Claret, ...	10
Cider, whole, . . .	9	Malaga, ...	16
Ale, ...	8	Champagne, ...	14
Porter, ...	7	Burgundy, ...	12
Brown Stout, ...	6	Muscat, ...	17
Perry, ...	9	Currant, ...	19

Proof of Spirituous Liquors.

The weight, in air, of a cubic inch of *Proof Spirits*, at 60° F., is 233 grains; therefore, an inch cube of any heavy body, at that temperature, weighing 233 grains less in spirits than in air, shows the spirits in which it is weighed to be *proof*. If the body lose less of its weight, the spirit is above proof, — if more, it is below.

Comparative Weight of different kinds of Timber in a green and perfectly seasoned state.

Assuming the weight of each kind destitute of water to be 100, that of the same kind green is as follows: —

Ash,	153	Cedar,	148	Maple, red,	149
Beech,	174	Elm, swamp,	198	Oak, Am.,	151
Birch,	169	Fir, Amer.,	171	Pine, white,	152

NOTE. — Woods which have been felled, cleft and housed for 12 months, still retain from 20 to 25 per cent. of water. They therefore contain but from 75 to 80 per cent. of heating matter; and it will require from 23 to 29 per cent. the weight of such woods to dispel the water they contain. They are, therefore, less valuable by weight, as fuel, by this per cent., than woods perfectly free from moisture. They never, however, contain, exposed to an ordinary atmosphere, less than 10 per cent. of water, however long kept; and even though rendered anhydrous by a strong heat, they again imbibe, on exposure to the atmosphere, from 10 to 12 per cent. of dampness.

Relative power of different seasoned Woods, Coals, &c., as fuel, to produce heat, — the Woods supposed to be seasoned to mean dryness, (77½ per cent.,) and the other articles to contain but their usual quantity of moisture.

	Ratio of Heating Power per equal Bulk.	Ratio of Heating Power per equal Weight.
Hickory, shell-bark,	1.00	1.00
" red-heart,	.81	.99
Walnut, com.	.95	.98
Beech, red,	.74	.99
Chestnut,	.49	.98
Elm, white,	.58	.98
Maple, hard,	.66	.98
Oak, white,	.81	.99
" red,	.69	.99
Pine, white,	.42	1.01
" yellow,	.48	1.03
Birch, black,	.63	.99
" white,	.48	.99
Coal, Cumberland, (bit.)	2.56	2.28
" Lackawanna (anth.)	2.28	2.22
" Lehigh, "	2.39	2.03
" Newcastle, (bit.)	2.10	1.96
" Pictou, (bit.)	2.21	1.91
" Pittsburgh, (bit.)	1.78	1.82
" Peach Mountain, (anth.)	2.69	2.29
Charcoal,	1.14	2.53
Coke, Virginia, natural,	1.89	2.12
" Cumberland,	1.31	2.25
Peat, ordinary,		62
Alcohol, common,		2.02
Beeswax, yellow,		2.90
Tallow,		3 10

NOTE. — By help of the preceding table, the price of either one article being known, the relative or par value of either other, as fuel, may be readily ascertained: — EXAMPLE:

Maple (66) : $5.00 : : Pine (42) : $3.18.

ILLUMINATION — ARTIFICIAL.

The following TABLE shows: —

1. The materials and methods of using — column *Materials.*

2. The comparative maximum intensity of light afforded by each material, used or consumed as indicated, — column *Intensities.*

3. The weight, in grains, of material consumed per hour, by each method respectively, in producing its respective light, or light of intensity ascribed — column *Weight.*

4. The *ratio* of weight required of each material, under each special method of consumption, for the production of equal lights in equal times — column *Ratios.*

Materials.		Intens.	Weight.	Ratios.
Camphene,	Paragon Lamp,	16.	853	1.
Sperm Oil,	Parker's heating Lamp,	11.	696	1.19
" "	Mech. or Carcel "	10.	815	1.53
" "	French annular "	5.	543	2.04
" "	Common hand "	1.	112	2.10
Whale "	p'f'd., P's heating "	9.	780	1.63
Wax Candles,	3's or 4's, 15 in. or 12 in.,	1.	125	2.35
" "	6's, 9 in.,	.92	122	2.50
Sperm "	4's, 13½ in.,	1.	142	2.66
Stearine "	4's, 13½ in.,	1.	168	3.15
Tallow, "	dipped, 10's,	.70	150	4.02
" "	mould, 10's,	.66	132	3.75
" "	" 8's,	.57	132	4.35
" "	" 6's,	.79	163	3.87
" "	" 4's, 13¾ in.,	1.	186	3.49
" *Coal Gas,*" intensity being		1.	411	

NOTE. — The consumption of 1.43 cubic feet of gas per hour, gives a light equal to one wax candle, — the consumption of 1.96 cubic feet per hour, a light equal to four wax candles, and the consumption of 3 cubic feet per hour, a light equal to ten wax candles. A cubic foot of gas weighs 286 grains.

The average yield of bi-carbureted hydrogen — Olefiant gas — Coal Gas, obtained from the following articles, is as annexed.

1 lb. Bituminous Coal,	4½ cubic feet.
1 lb. Oil, or Oleine,	15 " "
1 lb. Tar,	12 " "
1 lb. Rosin, or Pitch,	10 " "

A pound of good Lancashire cannel coal, or of good Scotch cannel, will yield, on an average, 5.95 cubic feet of good illuminating gas.

From the English Boghead cannel, by White's hydro-carbon process, 17 cubic feet to the pound are commonly obtained.

A pipe whose interior diameter is ½ inch, will supply gas equal in illuminating power to 20 wax candles.

The sp. gr. of coal-gas from cannel coal ranges from 0.640 to 0.370, while that from good working bituminous coal in general ranges from 0.420 to 0.370.

Results of Experiments by Mr. Clegg of London, relative to the conveyance of coal-gas through pipes of different lengths, diameters, &c.

Diameter of pipe in inches.	Length of pipe in yards.	Pressure in inches of water.	Specific gravity, air as 1.	Quantity in cubic feet discharged per hour.
0.5	10	1.25	0.4	120
0.5	59	1.25	0.4	60
0.62	41	1.38	0.559	99
0.62	62	1.34	0.559	83
0.62	93	1.34	0.559	74
0.62	119	1.34	0.559	57
0.62	138	1.34	0.559	53
2.	25	0.5	0.528	1630
10.	100	3.	0.4	120,000
10.	1760	3.	0.4	30,000
18.	1760	1.	0.4	66,000
26.	4300	0.475	0.42	80,000
26.	3130	0.8	0.42	103,000
26.	4300	2.25	0.42	175,000

From these experiments and others, Mr. Clegg derives the data for the following general approximate rules, viz.:

$Q = 1350d^2\sqrt{\frac{hd}{s(l+d)}}$; or, when the length of the pipe is above 400 or 500 times its diameter; or, in any case, if the pressure be measured in the pipe instead of in the gas-holder, $Q = 1350d^2\sqrt{\frac{hd}{ls}}$; in which Q represents the quantity of gas in cubic feet passed per hour, l the length of the pipe in yards, d the diameter of the pipe in inches, h the pressure in inches of water, and s the specific gravity of the gas.

Therefore, ordinarily for short mains of large diameters, or when the quotient of $36l \div d$ is less than 400,

$$l = \frac{1350^2 d^5 h}{Q^2 s} - d;\ h = \frac{Q^2 s(l+d)}{1350^2 d^5};\ \text{and } l + d = \frac{1350^2 d^5 h}{Q^2 s}.$$

And for long pipes of small diameters, or when l and $l + d$ are practically equal.

$$l = \frac{1350^2 d^5 h}{Q^2 s};\ h = \frac{Q^2 ls}{1350^2 d^5};\ \text{and } d = \sqrt[5]{\frac{Q^2 ls}{1350^2 h}}.$$

Now, in these formulas for Q, by Mr. Clegg, the probable retardation of the flow of the gas due to the friction of its particles in the pipes has been taken into account; and the results by the formulas agree as closely as could be expected with the given experiments, although the former average about 7 per cent. more than the latter.

The pipes are here supposed to be straight, and to lie horizontal, or equal in effect to that condition, and the gas is supposed to be delivered without force at the discharging end.

Example. — What pressure in inches of water is required to convey 852 cubic feet of gas per hour, of specific gravity 0.398, through a pipe 4 inches in diameter and 6 miles in length?

$$\frac{852^2 \times 10560 \times .398}{1350^2 \times 4^5} = 1.6348 \text{ inches. } Ans.$$

Note. — The illuminating power of coal-gas is nearly as its specific gravity, the more dense being the better.

THERMOMETERS.

	Boiling point.	Freezing point.
Fahrenheit's, ...	212°	32°
Reaumur's, ...	80°	0°
Centigrade, ...	100°	0°

To reduce Reaumur to Fahrenheit.

When it is desired to reduce the +°, (degrees above the zero): —

Rule. — Multiply the degrees Reaumur, by 2.25, and add 32° to the product; the sum will be the degrees Fahrenheit.

When it is desired to reduce the —°, (degrees below the zero): —

Rule. — Multiply the —° Reaumur by 2.25, and subtract the product from 32°; the difference will be the degrees Fahrenheit.

Example. — The degrees R. are 40; required the equivalent degrees F.

$$40 \times 2.25 = 90 + 32 = 122°. \quad Ans.$$

Example. — The degrees below 0, R., are 10; what are the corresponding degrees F.?

$$10 \times 2.25 = 22.5, \text{ and } 32 - 22.5 = 9\tfrac{1}{2}°. \quad Ans.$$

Example. — The degrees below 0, R., are 16; what point on the scale F. corresponds thereto?

$$16 \times 2.25 = 36, \text{ and } 32 - 36 = -4;\ 4° \text{ below } 0. \quad Ans.$$

To reduce the Centigrade to Fahrenheit.

Rule. — Multiply the degrees C. by 1.8, and in all other respects proceed as directed for Reaumur, above.

Note. — The *zero* of Wedgewood's pyrometer is fixed at the temperature of iron red-hot in daylight, = 1077° F., and each degree W. equals 130° F. The instrument is not considered reliable, and is but little used

HORSE POWER.

A HORSE-POWER, in machinery, as a measure of force, is estimated equal to the raising of 33000 lbs. over a single pulley one foot a minute, = 550 lbs. raised one foot a second, = 1000 lbs. raised 33 feet a minute.

ANIMAL POWER.

A man of ordinary strength is supposed capable of exerting a force of 30 lbs. for 10 hours in a day, at a velocity of 2½ feet a second, = 75 lbs. raised 1 foot a second.

The ordinary working power of a horse is calculated at 750 lbs. for 8 hours in a day, at a velocity of 2 feet a second, = 375 lbs. raised 1 foot a second, = 5 times the effective power of a man during associated labor, and 4 times his power per day; and as machinery may be supposed to work continually, = a trifle less than 23 per cent. per day of a machine horse-power.

STEAM.

Table exhibiting the expansive force and various conditions of steam under different degrees of temperature.

Degrees of heat.	Pressure in atmospheres.	Density. Water as 1.	Volume. Water as 1.	Spec. gravity. Air as 1.	Weight of a cubic foot in grains.
212	1	.00059	1694	.484	254
250.5	2	.00110	909	.915	483
276	3	.00160	625	1.330	700
293.8	4	.00210	476	1.728	910
308	5	.00258	387	2.120	1110
359	10	.00492	203	3.970	2100
418.5	20	.00973	106	7.440	3940

[An atmosphere is $14\frac{7}{10}$ lbs. to the square inch.]

NOTE. — By the above table it is seen that any given quantity of steam having a temperature of 212° F., occupies a space, under the ordinary pressure of the atmosphere, 1694 times greater than it occupied when as water in a natural state. It exerts a mechanical force, consequently, = 1694 times the weight or force of the atmosphere resting on the bulk from which it was generated, or resting on 1-1694th of the space it occupies. A force, if we consider the volume as so many cubic inches, equal to the raising of 2087 lbs. 12 inches high, by a quantity of steam less than a cubic foot, heated only to the temperature of boiling water, and weighing but 248 grains, and that, too, the product of a single cubic inch of water.

The mean pressure of the atmosphere at the earth's surface is equal to the weight of a column of mercury 29.9 inches in height, or to a column of water 33.87 feet in height, = 2116.8 lbs. per square foot, or

14.7 lbs. per square inch. Its density above the earth is uniformly less as its altitude is greater, and its extent is not above 50 miles — its mean altitude is about 45 miles; at 44 miles it ceases to reflect light. Were it of uniform density throughout, and of that at the surface, its altitude would be but 5¼ miles. Its weight is to pure water of equal temperature and volume, as 1 to 829. It revolves with the earth, and its average humidity, at 40° of latitude, is 4 grains per cubic foot. Its weight at 60°, b. 30, compared with an equal bulk of pure water at 40°, b. 30, is as 1 to 830.1.

VELOCITY AND FORCE OF WIND.

Appellations.	Mean velocity in Miles per hour.	Mean velocity in Feet per second.	Force in lbs. per square foot.
Just perceptible, . .	2½	3⅔	.032
Gentle, pleasant wind, . .	4½	6⅔	.101
Pleasant, brisk gale, . .	12½	18⅓	.80
Very brisk, " . .	22½	33	2.52
High wind, . . .	32½	47⅔	5.23
Very high wind, . . .	42½	62⅓	8.92
Storm, or tempest, . .	50	73⅓	12.30
Great storm, . . .	60	88	17.71
Hurricane, . . .	80	117⅓	31.49
Tornado, moving buildings, &c.,	100	146.7	49.20

The curvature of the earth is 6.99 inches (.5825 foot) in a single statute mile, or 8.05 inches in a geographical mile, and is as the square of the distance for any distance greater or less, or space between two levels; thus, for three statute miles it is

$$1 : 3^2 :: 6.99 : 5\frac{1}{4} \text{ feet, nearly.}$$

The horizontal refraction is $\frac{1}{13}$.

Degrees of longitude are to each other in length, as the cosines of their latitudes. At the equator a degree of longitude is 60 geographical miles in length, at 90° of latitude it is 0; consequently, a degree of longitude at

5° . .	= 59.77 miles.	40° . .	= 45.96 miles.
10° . .	= 59.09 "	50° . .	= 38.57 "
20° . .	= 56.38 "	70° . .	= 20.52 "
30° . .	= 51.96 "	85° . .	= 5.23 "

Time is to longitude 4 minutes to a degree, — faster, east of any given point; slower, west.

The mean velocity of sound at the temperature of 33° is 1100 feet a second. Its velocity is increased ½ a foot a second for every degree above 33°, and decreased $\frac{1}{2}$ a foot a second for every degree below 33 .

In water, sound passes at the rate of 4,708 feet a second.

Light travels at the rate of 192,000 miles per second.

GRAVITATION.

GRAVITY, or GRAVITATION, is a property of all bodies, by which they mutually attract each other proportionally to their masses, and inversely as the square of the distance of their centres apart. Practically, therefore, with reference to our Earth and the bodies upon or near its surface, gravity is a constant force centred at the Earth's centre, and is there continually operating to draw all bodies with a uniformly accelerating velocity to that point, and through very nearly equal spaces, in equal intervals of time from rest, at all localities.

Putting R' to represent the Equatorial radius of the earth, and r to represent the Polar, and making $R' = 3962.5$ statute miles, and $r = 3949.5$, which is nearly in accordance with the mean of the most reliable measurements of the arcs of a degree of latitude at different localities, we have $\varepsilon^2 = (R'^2 - r^2) \div R'^2 = .006550751$, the square of the ellipticity of the earth, and $R = 2R' \div (2 + \varepsilon^2 \sin^2 l)$, the radius at any given latitude l.

And since the initial velocity due to gravity at the level of the sea at the Equator is $G = 32.0741$ feet per second, or, in other words, since a body falling in vacuo at the equator, at the level of the sea, describes a space of 16.03705 feet in the first second of time from rest, we have $g = [R' \sqrt{G} \div R]^2$, the initial velocity at the level of the sea at any given radius R; or $g = (22441.2 \div R)^2$.

And finally $g = \left(\frac{22441.2}{R}\right)^2 \times \left(1 - \frac{2h}{5280R}\right)$ at any given radius R, at any given altitude, h, in feet, above the level of the sea.

NOTE.—When l, reckoned from the equator, is higher than 45°, $\sin^2 l = \cos^2 (90 - l)$.

The *Impulse*, or force with which a falling body strikes, is the product of its weight and velocity (the weight multiplied by the square root of the product of the space fallen through and 64.33, or 4 times $16\frac{1}{12}$); thus, 100 lbs., falling 50 feet, will strike with a force,

$$50 \times 64.333 = \sqrt{3216.66} = 56.71 \times 100 = 5671 \text{ lbs.}$$

An entire revolution of the earth, from west to east, is performed in 23 hours, 56 minutes, and 4 seconds. A solar year = 365 days, 5 hours, 48 minutes, 57 seconds.

The area of the earth is nearly 197,000,000 square miles. Its crust is supposed to be about 30 miles in thickness, and its mean density 5 times that of water. About $\frac{3}{4}$ of its area, or 150,000,000 square miles, is covered by water. The portions of land in the several divisions, in square miles, are, in round numbers, as follows, viz: —

Asia, . .	16,300,000	Europe, . .	3,700,000
Africa, . .	11,000,000	Australia, . .	3,000,000
America, . .	14,500,000		

America is 9000 miles long, or $\frac{36}{100}$ the circumference of the earth.

The population of the globe is about 1,000,000,000, of which there are, in

Asia, . .	456,000,000	Africa, . .	62,000,000
Europe, . .	258,000,000	America, . .	55,000,000

CHEMICAL ELEMENTS.

The chemical elements — simple substances in nature — as far as has been determined, are 62 in number: 13 non-metallic and 49 metallic.

Of the non-metallic, 5 — *bromine, chlorine, fluorine, iodine,* and *oxygen,* (formerly termed "*supporters of combustion,*") have an intense affinity for all the others, which they penetrate, corrode, and apparently consume, always with the production, to some extent, of light and heat. They are all non-conductors of electricity and negative electrics.

The remaining 8 — *hydrogen, nitrogen* or *azote, carbon, boron, silicon, phosphorus, selenium,* and *sulphur,* are eminently susceptible of the impressions of the preceding five; when acted upon by either of them to a certain extent, light and heat are manifestly evolved, and they are thereby converted into incombustible compounds.

Of the metals, 7 — *potassium, sodium, calcium, barytium, lithium, strontium,* and *magnesia,* by the action of oxygen, are converted into bodies possessed of *alkaline* properties.

Seven of them — *glucinum, erbium, terbium, yttrium, allumium, zirconium,* and *thorium,* — by the action of oxygen, are converted into the *earths* proper.

In short, all the metals are acted upon by oxygen, as also by most or all of the non-metallic family. The compounds thus formed are *alkaline, saline,* or *acidulous,* or an *alkali,* a *salt,* or an *acid,* according to the nature of the materials and the extent of combination.

Metals combine with each other, forming *alloys.* If one of the metals in combination is mercury, the compound is called an *amalgam.*

Silicon is the base of the mineral world, and *carbon* of the organized.

For a very general list of the metals, see TABLE OF SPECIFIC GRAVITIES.

TABLE

Exhibiting the Elementary Constituents and per cent. by weight of each, in 100 parts of different compounds.

Compounds.		Constituents and per cent.			
		Hydrogen.	Oxygen.	Nitrogen.	Carbon.
Atmospheric air, ...	*a*		19.96	79.84	
Water, pure, . . .		11.1	88.9		
Alcohol, anhydrous, . .		12.9	34.44		52.66
Olive oil, ...		13.4	9.4		77.2
Sperm "		10.97	10.13		78.9
Castor "		10.3	15.7		74.00
Stearine, (solid of fats,) .		11.23	6.3	0.30	82.17
Oleine, (liquid of fats,) .		11.54	12.07	0.35	76.03
Linseed oil, ...		11.35	12.64		76.01
Oil of turpentine, . .		11.74	3.66		84.6
"*Camphene*," (pure spts. turp.)		11.5			88.5
Caoutchouc, (gum elastic,) .		10.			90.
Camphor, ...		11.14	11.48		77.38
Copal, resin, . . .		9.	11.1		79.9
Guaiac, resin, . . .		7.05	25.07		67.88
Wax, yellow, . . .		11.37	7.94		80.69
Coals, cannel, . . .		3.93	21.05	2.80	72.22
" Cumberland, . .		3.02	14.42	2.56	80.
" Anthracite, . .	*b*				93.
Charcoal, ...					97.
Diamond, ...					100.
Oak wood, dry, . . .	*c*	5.69	41.78		52.53
Beech " "		5.82	42.73		51.45
Acetic acid, dry, . .		5.82	46.64		47.54
Citric " crystals, . .		4.5	59.7		35.8
Oxalic " dry, . .			79.67		20.33
Malic, " crystals, . .		3.51	55.02		41.47
Tartaric " dry, . .		3.	60.2		36.80
Formic " " . . .		2.68	64.78		32.54
Tannin, tannic acid, solid, .		4.20	44.24		51.56
Nitric acid, dry, . . .			73.85	26.15	
Nitrous " anhydrous, liquid,			61 32	30.68	
Ammoniacal gas, . .		17.47		82.53	
Carbonic acid " . . .			72.32		27.68
Carb. hydrogen gas, . .		24.51			75.49
Bi-carb. hyd., olefient gas, .		14.05			85.95
Cyanogen " .				53.8	46.2
Nitric oxyde " .			53.	47.00	
Nitrous " " .			36.36	63.64	
Ether, sulphuric, . . .		13.85	21.24		65.05
Creosote, ...		7.8	16.		76.2

Compounds.		Constituents and per cent. Hydrogen.	Oxygen.	Azote.	Carbon.
Morphia,		6.37	16.29	5.	72.34
Quina, — quinine,		7.52	8.61	8.11	75.76
Veratrine,		8.55	19.61	5.05	66.79
Indigo,		4.38	14.25	10.	71.37
Silk, pure white,		3.94	34.04	11.33	50.69
Starch, — farina, dextrine,		6.8	49.7		43.5
Sugar,		6.29	50.33		43.38
Gluten,		7.8	22.	14.5	55.7
Wheat,	*c*	6.	44.4	2.4	47.02
Rye,		5.7	45.3	1.7	47.03
Oats,		6.6	38.2	2.3	52.9
Potatoes,		6.1	46.4	1.06	45.9
Peas,		6.4	41.3	4.3	48.
Beet root,		6.2	46.3	1.8	45.7
Turnips,		6.	45.9	1.8	46.3
Fibrin,	*d*	7.03	20.30	19.31	53.36
Gelatin,	*d*	7.91	27.21	17.	47.88
Albumen,	*d*	7.54	23.88	15.70	52.88

Muriatic acid gas, — Hydrogen 5.53 + 94.47 chlorine.
Sulphuric acid, dry, — Oxygen 79.67 + 20.33 sulphur.
Silicic acid — Silica, dry, — Oxygen 51.96 + 48.04 silicon.
Boracic acid — Borax, dry, — " 68.81 + 31.19 boron.

a. The atmosphere, in addition to its constituents as given in the table, contains, besides a small quantity of vapor, from 1 to 3 parts in a thousand of carbonic acid gas, and a trace merely of ammoniacal gas.

b. Anthracite coal, charcoal, plumbago, coke, &c., have no other constituent than carbon; they are combined, to a small extent, with foreign matters, such as iron, silica, sulphur, alumina, &c.

c. The constituents of woods, grains, &c., are given per cent., without regard to the foreign matters (*metallic*) which they contain. In *oak*, *chestnut*, and *Norway pine*, the ashes amount to about $\frac{4}{10}$ of 1 per cent., and in *ash* and *maple* to $\frac{7}{10}$ of 1. In anthracite coals, at an average, they are about 7 per cent.

d. Fibrin, *Gelatin*, *Albumen* — Proximate animal constituents — Nutritious properties of animal matter.

Fibrin is the basis of the muscle (lean meat) of all animals, and is also a large constituent of the blood.

Gelatin exists largely in the skin, cartilages, ligaments, tendons and bones of animals. It also exists in the muscles and the membranes.

Albumen exists in the skin, glands and vessels, and in the serum of the blood. It constitutes nearly the whole of the white of an egg.

THE RELATIVE QUANTITIES BY VOLUME of the several gases going to constitute any particular compound, are readily ascertained by help of their respective specific gravities, compared with their relative weights, as given per cent. in the preceding table: — thus, the sp. gr. of hydrogen is .0689, and that of oxygen 1.1025, and 1.1025 ÷ .0689 = 16; showing the weight of the latter to be 16 times that of the former per equal volumes, or, relatively, as 16 to 1. The per cent. by weight, as shown by the table, in which these two gases combine to form water, for instance, is 11.1 and 88.9; or 11.1 of hydrogen and 88.9 of oxygen in 100 of the compound; or as 88.9 ÷ 11.1, — as 8 to 1: 16 ÷ 8 = 2: two volumes, therefore, of the lighter gas (hydrogen) combine with one of oxygen to form water. Water, consequently, is a Protoxide of Hydrogen.

Upon the principle of ATOMIC WEIGHTS, — relative quantities, by weight, in which the elements combine in forming compounds, based upon the standard already shown, — we have, for water, $H^1 + O^8 =$ Aq. 9. That is, an atom of hydrogen is represented by 1, an atom of oxygen by 8, and an atom of water by 9.

By the same rule as the preceding, the constituents of atmospheric air are found to be to each other, in volume, as 4 to 1; four volumes of nitrogen and one volume of oxygen make one volume of atmospheric air. The weight of nitrogen to hydrogen, per equal volumes, is as .972 to .0689, as 14.11 to 1. Atomically, therefore, it is as 7.055 to 1; hence, we have $N^4 + O = 36.22$, the atomic weight of atmospheric air.

The vast condensation of the gases which takes place, in some instances, in forming compounds, may be conceived of, and the process for determining the same exhibited by a single illustration. We will take, for example, water. A single cubic inch of distilled water, at 60°, weighs 252.48 grains. Its weight is to that of dry atmosphere, at the same temperature, as 827.8 to 1. A cubic inch of dry atmosphere, therefore, at that density, weighs .305 of a grain. Hydrogen, we find by the table of Specific Gravities, weighs .0689 as much as atmosphere, and oxygen 1.1025 as much. A cubic inch of hydrogen, therefore, weighs .0689 × .305 = .0210145 of a grain, and a cubic inch of oxygen 1.1025 × .305, = .3362625 of a grain. The constituents of water by volume are 2 of the first mentioned gas to 1 of the latter; and .0210145 × 2 + .3362625 = .3782915 of a grain, = weight of three cubic inches of the uncondensed compound, ⅓ of which, .1260972 of a grain, is the weight of a volume 1 cubic inch.

As the weight of a given volume of the uncondensed compound, is to the weight of an equal volume of the condensed compound, so are their respective volumes, inversely: then —

.1260972 : 252.48 :: 1 : 2002.26, the number of cubic inches of the two gases condensed into 1 inch to form water; a condensation of 2001 times. Of this volume of gases, ⅔, or 1334.84 cubic inches, is hydrogen; the remaining third, 667.42 cubic inches, is oxygen.

The foregoing method, though strictly correct, does not exhibit in a general way the most expeditious for solving questions of that nature, the condensation which takes place in the gases on being converted into solids, or dense compounds. It was resorted to, in part, as a means through which to exhibit principles and proportions pertaining thereto.

As before; one cubic inch of water weighs 252.48 grains, $\frac{1}{9}$ of which, or 28.05+ grains, is hydrogen, and $\frac{8}{9}$, or 224.43— grains, is oxygen. The volume of 1 grain of oxygen is 2.97+ cubic inches, and the volume of hydrogen is 16 times as much, or 47.58+ cubic inches. Therefore, $28.05 \times 47.58 = 1334.62$, and $224.43 \times 2.97 = 665.56$, $= 2001.18$, condensation, as before.

Properties of the SIMPLE SUBSTANCES, *and some of their compounds, not given in the foregoing.*

BROMINE, — at common temperatures, a deep reddish-brown volatile liquid; taste caustic; odor rank; boils at 116°; congeals at 4°; exists in sea-water, in many salt and mineral springs, and in most marine plants; action upon the animal system very energetic and poisonous — a single drop placed upon the beak of a bird destroys the bird almost instantly. A lighted taper, enveloped in its fumes, burns with a flame green at the base and red at the top; powdered tin or antimony brought in contact is instantly inflamed; potash is exploded with violence.

CHLORINE, — a greenish-yellow, dense gas; taste astringent; odor pungent and disagreeable; by a pressure of 60 lbs. to the square inch is reduced to a liquid, and thence, by a reduction of the temperature below 32°, into a solid. It exists largely in sea-water — is a constituent of common salt, and forms compounds with many minerals; is deleterious, irritating to the lungs, and corrosive; has eminent bleaching properties, and is the greatest disinfecting agent known; a lighted taper immersed in it burns with a red flame; pulverized antimony is inflamed on coming in contact, so is linen saturated with oil of turpentine; phosphorus is ignited by it, and burns, while immersed, with a pale-green flame; with hydrogen, mixed measure for measure, it is highly explosive and dangerous.

FLUORINE, — a gas, similar to chlorine, — exists abundantly in *fluor-spar*.

OXYGEN, — a transparent, colorless, tasteless, inodorous, innoxious gas; supports respiration and combustion, but will not sustain life for any length of time, if breathed in a pure state. It is by far the most abundant substance in existence; constitutes $\frac{1}{5}$ of the atmosphere;

$\frac{8}{9}$ of water; and nearly the whole crust of the earth is oxidized substances. For further combinations and properties, see tables of *Elementary Constituents* and *Chemical Elements.*

IODINE, — at common temperatures, a soft, pliable, opaque, bluish-black solid; taste acrid; odor pungent and unpleasant; fuses at 225°; boils at 347°; its vapor is of a beautiful violet color; it inflames phosphorus, and is an energetic poison; exists mainly in sea-weeds and sponges.

HYDROGEN, — a transparent, colorless, tasteless, inodorous, innoxious gas; if pure, will not support respiration; if mixed with oxygen, produces a profound sleep; exists largely in water; is the basis of most liquids, and is by far the lightest substance known; burns in the atmosphere with a pale, bluish light; mixed with common air, 1 measure to 3, it is explosive; mixed with oxygen, 2 measures to 1, it is violently so.

NITROGEN, or *Azote,* — a transparent, colorless, tasteless, inodorous gas; will not support respiration or combustion, if pure; exists largely as a constituent of the atmosphere — in animals, and in fungous plants; is evolved from some hot springs; in connection with some bodies, appears combustible.

CARBON, — the *diamond* is the only pure carbon in existence; pure carbon cannot be formed by art; *charcoal* is 97 per cent. carbon; *plumbago*, 95; *anthracite*, 93. Carbon is supposed by some to be the *hardest* substance in nature. A piece of charcoal will scratch glass; but it is doubtful if this is not due to the form of its crystals, rather than to the first mentioned quality. It is doubtless the most *durable.* For combinations, &c., see table.

BORON, — a tasteless, inodorous, dark olive-colored solid.

SILICON, — a tasteless, inodorous solid, of a dark-brown color; exists largely in soils, quartz, flint, rock-crystal, &c.; burns readily in air — vividly in oxygen gas; explodes with soda, potassa, barryta.

PHOSPHORUS, — a transparent, nearly colorless solid, of a wax-like texture; fuses at 108°, and at 550° is converted into a vapor; exists mainly in bones — most abundant in those of man — is poisonous; at common temperatures it is luminous in the dark, and by friction is instantly ignited, burning with an intense, hot, white flame; must be kept immersed in water.

SELENIUM, — a tasteless, inodorous, opaque, brittle, lead-colored solid, in the mass; in powder, a deep-red color; becomes fluid at 216°, boils at 650°; vapor, a deep yellow; exists but sparingly, mainly in combination with volcanic matter; is found in small quantities combined with the ores of lead, silver, copper, mercury.

Ammoniacal gas, — $N + H^3$; transparent, colorless, highly pungent and stimulating; alkaline; is converted into a transparent liquid by a pressure of 6.5 atmospheres, at 50°; does not support respiration; is inflammable.

Carbonic acid gas, — $C + O^2$; transparent, colorless, inodorous, dense; is converted into a liquid by a pressure of 36 atmospheres; exists extensively in nature, in mines, deep wells, pits; is evolved from the earth, from ordinary combustion, especially from the combustion of charcoal, and from many mineral springs; is expired by man and animals; forms 44 per cent. of the *carbonate of lime* called marble; the brisk, sparkling appearance of soda-water, and most mineral waters, is due to its presence. It is neither a combustible nor a supporter of combustion; and, when mixed with the atmosphere to an extent in which a candle will not burn, is destructive of life. Being heavier than atmosphere, it may be drawn up from wells in large open buckets; or it may be expelled by exploding gunpowder near the bottom. Large quantities of water thrown in will absorb it.

The above gas is expired by man to the extent of 1632 cubic inches per hour; it is generated by the burning of a wax candle to the extent of 800 cubic inches per hour; and, by the burning of "*Camphene*," (in the production of light equal to that afforded by 1 wax candle,) to the extent of 875 cubic inches per hour. Two burning candles, therefore, vitiate the air to about the same extent as 1 person.

Carbonic oxide gas, — $C + O$; transparent, colorless, insipid; odor offensive; does not support combustion; an animal confined in it soon dies; is highly inflammable, burning with a pale blue flame; mixed with oxygen, 1 to 2, is explosive — with atmosphere, even in small quantity, is productive of giddiness and fainting.

Carbureted hydrogen gas, — $C + H^2$; transparent, colorless, tasteless, nearly inodorous; exists in marshes and stagnant pools — is there formed by the decomposition of vegetable matter; extinguishes all burning bodies, but at the same time is itself highly combustible, burning with a bright but yellowish flame; it is destructive to life, if respired.

Cyanogen — Bicarburet of Nitrogen — a *gas*, — $N + C^2$; transparent, colorless, highly pungent and irritating; under a pressure of

3.6 atmospheres, becomes a limpid liquid; burns with a beautiful purple flame.

Hydrochloric acid gas — Muriatic acid gas, — H + Cl. (chlorine); transparent, colorless, pungent, acrid, suffocating; strong acid taste.

Nitrous oxide gas — Protoxide of Nitrogen, "laughing gas," — N + O; transparent, colorless, inodorous; taste sweetish; powerful stimulant, when breathed, exciting both to mental and muscular action; can support respiration but from 3 to 4 minutes; is often pernicious in its effects.

Nitric oxide gas — Binoxide of Nitrogen, — N + O^2; transparent, colorless; wholly irrespirable; lighted charcoal and phosphorus burn in it with increased brilliancy.

Olefiant gas — Bicarburcted hydrogen gas — "coal gas," — C^2 + H^2; transparent, colorless, tasteless, nearly inodorous, when pure; does not support respiration or combustion; a lighted taper immersed in it is immediately extinguished. It burns with a strong, clear, white light; mixed with oxygen, in the proportion of 1 volume to 3, it is highly explosive and dangerous.

Phosphureted hydrogen gas, — P + H^3; colorless; odor highly offensive; taste bitter; exists in the vicinity of swamps, marshes, and grave-yards; is formed by the decomposition of bones, mainly; is highly inflammable; takes fire spontaneously on coming in contact with the atmosphere; mixed with pure oxygen, it explodes. It is the veritable "Will o' the wisp."

Sulphurcted hydrogen gas — Hydrosulphuric acid gas, — S + H; transparent, colorless; taste exceedingly nauseous; odor offensive and disgusting; is furnished by the sulphurets of the metals in general — also by filthy sewers and putrescent eggs. It is very destructive to life; placed on the skin of animals, it proves fatal. It burns with a pale blue flame, and, mixed with pure oxygen, it is explosive.

Hydrocyanic acid — Prussic acid, — N + C^2 + H; a colorless, limpid, highly volatile liquid; odor strong, but agreeable — similar to that of peach-blossoms; it boils at 79° and congeals at 0; exists in laurel, the bitter almond, peach and peach kernel. It is a most virulent poison, — a drop placed upon a man's arm caused death in a few minutes. A cat, or dog, punctured in the tongue with a needle fresh dipped in it, is almost instantly deprived of life.

Hydrofluoric acid, — F + H; a colorless liquid, in well stopped lead or silver bottles, at any temperature between 32° and 59°. It is obtained by the action of sulphuric acid on fluor-spar. It readily acts upon and is used for etching on glass. It is the most destructive to animal matter of any known substance.

Nitrohydrochloric acid — "*aqua regia*," — (1 part nitric acid and 4 parts muriatic acid, by measure ;) — a solvent for gold. The best solvent for gold is a solution of sal ammoniac in nitric acid.

Nitrosulphuric acid, — (1 part nitric acid and 10 parts sulphuric acid, by measure) — a solvent for silver ; scarcely acts upon gold, iron, copper, or lead, unless diluted with water ; is used for separating the silver from old plated ware, &c. The best solvent for silver, and one which will not act in the least upon gold, copper, iron, or lead, is a solution of 1 part of nitre in 10 parts of concentrated sulphuric acid, by weight, heated to 160°. This mixture will dissolve about $\frac{1}{6}$ its weight of silver. The silver may be recovered by adding common salt to the solution, and the chloride decomposed by the carbonate of soda.

Selenic acid, — $Se + O^3$; obtained by fusing nitrate of potassa with selenium — a solvent for gold, iron, copper, and zinc.

Silicic acid, — (*Silica* — silex ; base *Silicon*) — $Si + O^3$; exists largely in sand. Common glass is fused sand and protoxide of potassium (carbonate of potassa — *potash*) in the proportion of 1 part by weight of the former to 3 of the latter.

Manganese, compounded with oxygen, in different proportions, imparts the various colors and tints given to fancy glass ware, now so generally in vogue.

Butylene, — $C^4 + H^4$; sp. gr. 1.9348 ; a gaseous hydro-carbon derived from the distillation of coal-tar ; illuminating power, compared with that of olefient gas, as 2 to 1.

Propylene, — $C^3 + H^3$; sp. gr. 1.4511 ; a gaseous hydro-carbon derived from the distillation of coal-tar ; illuminating power, compared with that of olefient gas, as 1.5 to 1.

Napthaline Vapor, — $C^{10} + H^4$, the vapor of solidified olefient gas.

Turpentine Vapor, — $C^{10} + H^8$.

SECTION III.

PRACTICAL ARITHMETIC.

VULGAR FRACTIONS.

A fraction is one or more parts of a UNIT.

A vulgar fraction consists of two *terms*, one written above the other, with a line drawn between them.

The term below the line is called the *denominator*, as showing the denomination of the fraction, or number of parts into which the unit is broken.

The term above the line is called the *numerator*, as numbering the parts employed. These together constitute the fraction and its value.

A vulgar fraction always denotes *division*, of which the denominator is the *divisor* and the numerator the *dividend*. Its value as a *unit* is the quotient arising therefrom.

A simple fraction is either a proper or improper fraction.

A proper fraction is one whose numerator is less than its denominator, as $\frac{1}{2}$, $\frac{2}{3}$, $\frac{21}{50}$, &c.

An improper fraction has its numerator equal to or greater than its denominator, as $\frac{2}{2}$, $\frac{4}{3}$, $\frac{24}{15}$, &c.

A mixed fraction is a compound of a whole number and a fraction, as $1\frac{1}{2}$, $5\frac{11}{32}$, $12\frac{3}{16}$, &c.

A compound fraction is a fraction of a fraction, as $\frac{1}{2}$ of $\frac{3}{8}$; $\frac{3}{4}$ of $\frac{8}{5}$ of $\frac{12}{17}$, &c.

A complex fraction has a fraction for its numerator or denominator, or both, as $\frac{\frac{1}{2}}{3}$, $\frac{4}{\frac{3}{5}}$, $\frac{\frac{1}{2}}{\frac{3}{4}}$, $\frac{5\frac{1}{2}}{4}$, &c., and is read $\frac{1}{2} \div 3$; $4 \div \frac{3}{5}$; $\frac{1}{2} \div \frac{3}{4}$; $5\frac{1}{2} \div 4$, &c.

REDUCTION OF VULGAR FRACTIONS.

To reduce a fraction to its lowest terms.

This consists in concentrating the expression without changing the value of the fraction or the relation of its parts.

It supposes division, and, consequently, by a measure or measures common to both terms.

It is said to be accomplished when no number greater than 1 will divide both terms without a remainder: — therefore,

RULE. — Divide both terms by any number that will divide them without a remainder, and the quotient again as before; continue so to do until no number greater than 1 will divide them, — or divide by the greatest common measure at once.

EXAMPLE. — Reduce $\frac{864}{1512}$ to its lowest terms.

$$4\ )\ \frac{864}{1512} = \frac{216}{378} \div 2 = \frac{108}{189} \div 9 = \frac{12}{21} \div 3 = \frac{4}{7}. \quad \textit{Ans.}$$

To reduce an improper fraction to a mixed or whole number.

RULE. — Divide the numerator by the denominator and to the whole number in the quotient annex the remainder, if any, in form of a fraction, making the divisor the denominator as before; then reduce the fraction to its lowest terms.

EXAMPLE. $\frac{5}{4} = 1\frac{1}{4}$; $\frac{16}{12} = 1\frac{4}{12} = 1\frac{1}{3}$; $\frac{26}{13} = 2$.

To reduce a mixed fraction to an equivalent improper fraction.

RULE. — Multiply the whole number by the denominator of the fractional part, and to the product add the numerator, and place their sum over the said denominator.

EXAMPLE. — Reduce $3\frac{1}{4}$ and $12\frac{8}{9}$ to improper fractions.

$3 \times 4 = 12 + 1 = \frac{13}{4}$. *Ans.* $\qquad 12 \times 9 + 8 = \frac{116}{9}$. *Ans.*

To reduce a whole number to an equivalent fraction having a given denominator.

RULE. — Multiply the whole number by the given denominator, and place the said denominator under the product.

EXAMPLE. — How may 8 be converted into a fraction whose denominator is 12?

$8 \times 12 = \frac{96}{12}$. *Ans.*

To reduce a compound fraction to a simple one.

RULE. — Multiply all the numerators together for a numerator, and all the denominators together for a denominator; the fraction thus formed will be an equivalent, but often not in its lowest terms. Or, concentrate the expression, when practicable, by reciprocally expunging, or writing out, such factors as exist or are attainable common to both terms, and then multiply the remaining terms as directed above.

NOTE. — This last practice is called cancellation, or cancelling the terms. It consists, as has been stated, in reciprocally annulling, or casting out, equal values from both terms, whereby the expression is concentrated, and the relation of the parts kept undisturbed; and it may always be carried to the extent of reducing the fraction to its lowest terms, before any multiplication, as final, is resorted to; and often, therefore, to the extent that such multiplication is inadmissible, the terms having been cancelled by each other until but a single number is left in each.

EXAMPLE. — Reduce $\frac{2}{3}$ of $\frac{3}{4}$ of $\frac{1}{2}$ to a simple fraction.

Operation by multiplication, $\frac{2}{3} \times \frac{3}{4} \times \frac{1}{2} = \frac{6}{24} = \frac{1}{4}$. *Ans.*

Operation by cancellation, $\frac{\not{2}\ \not{3}\ 1}{\not{3}\ 4\ \not{2}} = \frac{1}{4}$. *Ans.*

EXAMPLE. — Reduce $\frac{2}{3}$ of $\frac{3}{4}$ of $\frac{12}{8}$ of $\frac{6}{8}$ of $\frac{5}{9}$ of 2 to a simple fraction.

By multiplication, $\frac{2}{3} \times \frac{3}{4} \times \frac{12}{8} \times \frac{6}{8} \times \frac{5}{9} \times \frac{2}{1} = \frac{4320}{6912} = \frac{5}{8}$. *Ans.*

The last example stated for cancellation, $\left\{ \frac{2\ \ 3\ \ 12\ \ 6\ \ 5\ \ 2}{3\ \ 4\ \ 8\ \ 8\ \ 9} \right.$

PROCESS OF CANCELLING THE ABOVE.

1. The 3 in num. equals the 3 in denom., therefore erase both.
2. The first 2 in num. equals or measures the 4 in denom. *twice*, therefore place a 2 under the 4, and erase the 4 and 2 which measured it — (as 4 : 2 : : 2 : 1.)
3. The 2 (remaining factor of 4 and 2 erased) in denom., and the remaining 2 in num., will cancel each other, — erase them.
4. The 12 and 6 in num. = 72, and the 9 and 8 in denom. = 72; these, therefore, in their relations as factors equal each other, and may be erased.

The remaining factors represent the true value of the compound fraction, and will be found = $\frac{5}{8}$, as by multiplication.

EXAMPLE. — Reduce $\frac{18}{13}$ of $\frac{7}{12}$ to a simple fraction.

$$\frac{\overset{\overset{3}{\not{9}}}{\not{1}\not{8}} \times 7}{13 \times \underset{\underset{2}{\not{6}}}{\not{1}\not{2}}}.\quad \text{Or, } \frac{\overset{3}{\not{1}\not{8}} \times 7}{13 \times \underset{2}{\not{1}\not{2}}}\ (= 18 \div 6, \text{ and } 12 \div 6) = \frac{3}{2} \times \frac{7}{13} = \frac{21}{26}.\ \textit{Ans.}$$

To reduce two or more fractions to a common denominator.

RULE.—Multiply each numerator by all the denominators except its own, for the new numerators; and multiply all the denominators together, for a common denominator.

NOTE.—Whole numbers and fractions other than simple, must first be reduced to simple fractions before they can be reduced to fractions having a common denominator.

EXAMPLE.—Reduce $\frac{2}{3}$ and $\frac{3}{4}$ to fractions having a common denominator.

$\frac{2}{3} + \frac{3}{4} = \frac{8}{12}$ and $\frac{9}{12}$; that is, $\frac{2}{3} = \frac{2 \times 4}{3 \times 4} = \frac{8}{12}$, and $\frac{3}{4} = \frac{3 \times 3}{4 \times 3} = \frac{9}{12}$; and $\frac{8}{12}$ and $\frac{9}{12}$ are fractions having a common denominator.

EXAMPLE.—Reduce $\frac{1}{2}$, $\frac{3}{5}$, $\frac{7}{8}$, and $\frac{14}{3}$ to fractions having a common denominator:

$\frac{1}{2} + \frac{3}{5} + \frac{7}{8} + \frac{14}{3} = \frac{120}{240}, \frac{144}{240}, \frac{210}{240}, \frac{1120}{240} = \frac{60}{120}, \frac{72}{120}, \frac{105}{120}, \frac{560}{120} = .5, .6, .875, 4\frac{2}{3}$.

To reduce a complex fraction to a simple one.

RULE. — Multiply the numerator of the upper fraction by the denominator of the lower, for the new numerator; and the denominator of the upper by the numerator of the lower for the new denominator.

EXAMPLES. — Reduce $\frac{\frac{1}{2}}{3}$, $\frac{4}{\frac{3}{5}}$, $\frac{\frac{1}{2}}{\frac{3}{4}}$, and $\frac{5\frac{1}{2}}{4}$, each to a simple fraction.

$\frac{1}{2} \div \frac{3}{1} = \frac{1}{6}$; $\frac{4}{1} \div \frac{3}{5} = \frac{20}{3}$; $\frac{1}{2} \div \frac{3}{4} = \frac{1}{2} \times \frac{4}{3}, = \frac{4}{6}, = \frac{2}{3}$; $5\frac{1}{2} = \frac{11}{2}$, and $\frac{11}{2} \times \frac{1}{4} = \frac{11}{8}, = 1\frac{3}{8}$. *Ans.*

To reduce Vulgar Fractions to equivalent Decimals.

RULE. — Divide the numerator by the denominator; the quotient is the decimal, or the whole number and decimal, as the case may be.

EXAMPLE. — Reduce $\frac{7}{8}$, $4\frac{3}{5}$, $\frac{14}{12}$, to decimals.

$7 \div 8 = 0.875$; $4\frac{3}{5} = \frac{23}{5}, = 4.6$; $14 \div 12 = 1.166+$. *Ans.*

To find the greatest common measure of two or more given numbers.

RULE. — Divide the given numbers by any measure common to them all, and set the quotients in a line beneath; then divide the quotients by any measure common to them, and set the quotients beneath; and so on until the quotients are no longer common multiples of any one number greater than unity; the product of all the divisors or common measures employed will be the greatest common measure.

EXAMPLE. — What is the greatest common measure of 84 and 36? Also of 32, 24, and 16? Also of 182, 104, and 52?

4)84.36	8)32.24.16	2)182.104.52
3)21 . 9	4 . 3 . 2.—8. *Ans.*	13) 91 . 52.26
7 . 3 — 4 × 3 = 12. *Ans.*		7 . 4 . 2.=26. *Ans.*

NOTE. — When any number in the series is prime to either of the others, the numbers are collectively incommensurable; that is to say, their greatest common measure is unity, or 1.

To find the least common multiple of two or more given numbers.

RULE. — Divide all the given numbers that are commensurable with each other by any measure that is common to them, and set the quotients, together with the undivided numbers, if any, in a line beneath; then divide the quantities in the second line as before, and so on until no two quantities in the last line are common multiples of any number greater than unity, or 1; the product of all the common measures employed into the product of all the numbers in the last line will be the least common multiple of the given numbers.

EXAMPLE. — What is the least common multiple of 27 and 36? Also of 182, 104, and 52? Also of 24, 14, 12, and 7?

9) 27 . 36
 3 . 4 $=4\times3\times9=108$. *Ans.*

2) 182 . 104 . 52
13) 91 . 52 . 26
2) 7 . 4 . 2
 7 . 2 . 1. $=728$. *Ans.*

7) 24 . 14 . 12 . 7
2) 24 . 2 . 12 . 1
6) 12 . 1 . 6
 2 . 1 . 1. 168. *Ans.*

ADDITION OF VULGAR FRACTIONS.

Sum of the products of each numerator with all the denominators except that of the numerator involved, forms numerator of sum.

Product of all the denominators forms denominator of sum.

RULE. — Arrange the several fractions to be added, one after another, in a line from left to right; then multiply the numerator of the first by the denominator of the second, and the denominator of the first by the numerator of the second, and add the two products together for the numerator of the sum; then multiply the two denominators together for its denominator; bring down the next fraction, and proceed in like manner as before, continuing so to do until all the fractions have been brought down and added. Or, reduce all to a common denominator, then add the numerators together for the numerator of the sum, and write the common denominator beneath.

EXAMPLES. — Add together $\frac{1}{2}$, $\frac{2}{3}$, $\frac{3}{4}$, and $\frac{4}{3}$.

$\frac{1}{2}+\frac{2}{3}=\frac{7}{6}+\frac{3}{4}=\frac{46}{24}+\frac{4}{3}=\frac{234}{72}=\frac{13}{4}=3\frac{1}{4}$. *Ans.*

$\frac{1}{2}=\frac{2}{4}+\frac{3}{4}=\frac{5}{4}=\frac{15}{12}$, and $\frac{2}{3}+\frac{4}{3}=\frac{6}{3}=\frac{24}{12}$, and $\frac{15}{12}+\frac{24}{12}=\frac{39}{12}$ $=\frac{13}{4}=3\frac{1}{4}$. *Ans.*

SUBTRACTION OF VULGAR FRACTIONS.

Product of numerator of minuend and denominator of subtrahend, forms numerator of minuend, for common denominator.

Product of numerator of subtrahend and denominator of minuend, forms numerator of subtrahend, for common denominator.

Product of denominators forms common denominator.

Difference of new found numerators forms the numerator, and common denominator the denominator, of the difference, or remainder sought.

RULE. — Write the subtrahend to the right of the minuend, with the sign (—) between them; then multiply the numerator of the minuend by the denominator of the subtrahend, and the denominator of the minuend by the numerator of the subtrahend; subtract the latter product from the former, and to the remainder or difference affix the product of the two denominators for a denominator; the sum thus formed is the answer, or true difference.

EXAMPLES. — Subtract $\frac{1}{2}$ from $\frac{3}{4}$, also $\frac{3}{5}$ from $\frac{11}{17}$.

$$\tfrac{3}{4} - \tfrac{1}{2} = \tfrac{6-4}{8}, = \tfrac{2}{8} = \tfrac{1}{4}. \quad \textit{Ans.}$$

$$\tfrac{11}{17} - \tfrac{3}{5} = \tfrac{55-51}{85} = \tfrac{4}{85}. \quad \textit{Ans.}$$

DIVISION OF VULGAR FRACTIONS.

Product of numerators of dividend and denominators of divisor, forms numerator of quotient.

Product of denominators of dividend and numerators of divisor, forms denominator of quotient; therefore,

RULE. — Write the divisor to the right of the dividend with the sign (÷) between them; then multiply the numerator of the dividend by the denominator of the divisor, for the numerator of the quotient, and the denominator of the dividend by the numerator of the divisor, for the denominator of the quotient. Or, invert the divisor, and multiply as in multiplication of fractions. Or, proceed by cancellation, when practicable.

EXAMPLES. — Divide $\frac{1}{2}$ by $\frac{3}{4}$; $\frac{3}{4}$ by $\frac{1}{2}$; $\frac{4}{3}$ by $\frac{11}{13}$; and $\frac{1}{2}$ of $\frac{3}{4}$ of $\frac{5}{6}$ of $\frac{4}{3}$ by $\frac{1}{4}$ of $\frac{1}{2}$ of $\frac{3}{4}$ of $\frac{2}{3}$.

$\frac{1}{2} \div \frac{3}{4} = \frac{4}{6}$; $\frac{3}{4} \div \frac{1}{2} = \frac{6}{4}$; $\frac{4}{3} \div \frac{11}{13} = \frac{52}{33}$; or $\frac{4}{3} \times \frac{13}{11} = \frac{52}{33}$. *Ans.*

$\frac{1}{2} \times \frac{3}{4} \times \frac{5}{6} \times \frac{4}{3} = \frac{60}{144} = \frac{5}{12}$, and $\frac{1}{4} \times \frac{1}{2} \times \frac{3}{4} \times \frac{2}{3} = \frac{6}{96} = \frac{1}{16}$, and $\frac{5}{12} \div \frac{1}{16} = \frac{80}{12} = \frac{20}{3} = 6\frac{2}{3}$. *Ans.*

FORM FOR CANCELLATION. — EXAMPLE LAST GIVEN.

$$\frac{1 \;\; 3 \;\; 5 \;\; 4 \quad 4 \;\; 2 \;\; 4 \;\; 3}{2 \;\; 4 \;\; 6 \;\; 3 \quad 1 \;\; 1 \;\; 3 \;\; 2} = \frac{20}{3}. \quad \textit{Ans., as above.}$$

NOTE. — The foregoing example can be cancelled to the extent of leaving but a 4 and a 5 (= 20) numerators, and a 3 denominator. Units, or 1's, in the expressions, are valueless, as a sum multiplied by 1 is not increased.

MULTIPLICATION OF VULGAR FRACTIONS.

Product of numerators of multiplier and multiplicand, forms numerator of product.

Product of denominators of multiplier and multiplicand, forms denominator of product.

RULE. — Multiply the numerators together for a numerator, and the denominators together for the denominator.

EXAMPLES. — Multiply $\frac{1}{2}$ by $\frac{1}{2}$; $\frac{3}{4}$ by 7; $\frac{11}{12}$ by $\frac{12}{7}$; $\frac{1}{2}$ of $\frac{2}{3}$ of $\frac{3}{4}$ by $\frac{3}{4}$ of $\frac{1}{2}$ of $\frac{2}{3}$.

$\frac{1}{2} \times \frac{1}{2} = \frac{1}{4}$; $\frac{3}{4} \times \frac{7}{1} = \frac{21}{4}$; $\frac{11}{12} \times \frac{12}{7} = \frac{132}{84} = \frac{11}{7}$; $\frac{1}{2} \times \frac{2}{3} \times \frac{3}{4} = \frac{6}{24} = \frac{1}{4}$, and $\frac{3}{4} \times \frac{1}{2} \times \frac{2}{3} = \frac{6}{24} = \frac{1}{4}$, and $\frac{1}{4} \times \frac{1}{4} = \frac{1}{16}$ *Ans.*

MULTIPLICATION AND DIVISION OF FRACTIONS, COMBINED.

It has been seen that a compound fraction is converted into an equivalent simple one, by multiplying the numerators together for a numerator, and the denominators together for a denominator; and it has also been seen that a series of simple fractions are converted into a product, by the same process. It is therefore evident that compound fractions and simple, or a series of compound and a series of simple, may be multiplied into each other, for a product, by multiplying all the numerators of both together for a numerator, and all the denominators of both together for a denominator; and that the product will be the same as would be obtained, if the compound were first converted into an equivalent simple fraction, and the simple fractions into a product or factor, and these multiplied together for a product.

It has also been seen that a fraction is divided by a fraction by multiplying the numerator of the dividend by the denominator of the divisor, for the numerator of the quotient, and the denominator of the dividend by the numerator of the divisor, for the denominator of the quotient; and that this multiplication becomes direct as in multiplying for a product, if the divisor is inverted. And it is clear that a compound divisor, or a series of simple divisors, or both, may be used instead of their simple equivalent, and with the same result, if all are inverted.

It is therefore evident that any proposition, or problem, in fractions, consisting of multiplications and divisions both, and these only, no matter how extensive and numerous, or whether in compound fractions, or simple, or both, may be solved, and the true result obtained, as a product, by simply multiplying all the numerators in the statement together for a numerator, and all the denominators in the statement for a denominator, all the divisors in the statement being inverted; that is, all the numerators of the divisors being made denominators in the statement, and all the denominators of the divisor being made numerators in the statement. And it is further evident that a proposition stated in this way, admits of easy cancellation as far as cancellation is practicable, which is often to great extent.

EXAMPLE.—It is required to divide 12 by $\frac{2}{3}$ of $\frac{3}{4}$; to multiply the quotient by the product of 4 and 8; to divide that product by $\frac{7}{2}$ of $\frac{3}{5}$ of 8; to multiply the quotient by $\frac{7}{8}$ of $\frac{8}{3}$ of $\frac{9}{14}$; and to divide that product by the product of 5 and 9.

STATEMENT.

(Dividends read from right to left, divisors from left to right.)

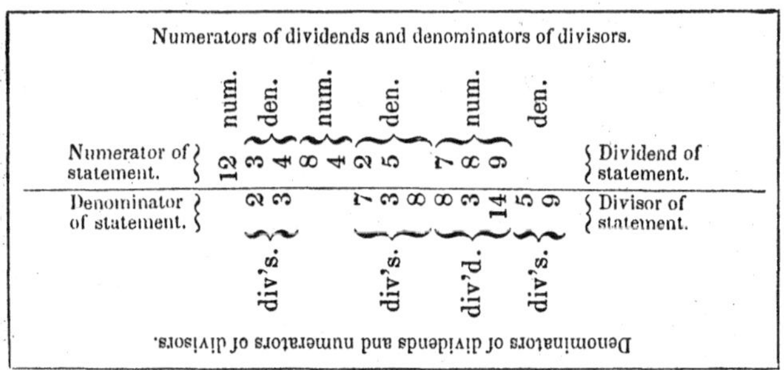

The *answer* to the above proposition is $1\frac{11}{21}$, and the proposition as stated may be readily cancelled to its lowest terms. It may be cancelled to the extent of leaving but 4, 4, 2 in the numerator, and 7, 3, in the denominator, $\frac{4\times4\times2}{7\times3}=\frac{32}{21}=1\frac{11}{21}$.

To reduce a fraction in a higher denomination to an equivalent fraction in a given lower denomination.

RULE. — Multiply the fraction to be reduced — numerators into numerator and denominators into denominator — by a fraction whose numerator represents the number of parts of the lower denomination, required to make ONE of the denomination to be reduced.

EXAMPLE. — Reduce $\frac{7}{8}$ of a foot to an equivalent fraction in inches.

$\frac{7}{8}\times\frac{12}{1}=\frac{84}{8}=\frac{21}{2}$. *Ans.*

EXAMPLE. — Reduce $\frac{5}{6}$ of a pound to an equivalent fraction in $\frac{2}{3}$ ounces.

$\frac{5}{6}\times\frac{16}{1}=\frac{80}{6}\div\frac{2}{3}=\frac{240}{12}=\frac{20}{1}$. *Ans.*

Or, $\frac{5}{6}\times\frac{16}{1}\times\frac{3}{2}=\frac{240}{12}=\frac{20}{1}$. *Ans.*

To reduce a fraction in a lower denomination to an equivalent fraction in a given higher denomination.

RULE. — Multiply the fraction to be reduced — numerator into denominator and denominator into numerator — by a fraction whose numerator represents the number of parts required of the lower denomination to make 1 of the higher.

EXAMPLE. — Reduce $\frac{21}{2}$ inches to an equivalent fraction in feet.

$\frac{21}{2}\div\frac{12}{1}=\frac{21}{24}=\frac{7}{8}$. *Ans.* Or, $\frac{21}{2}\times\frac{1}{12}=\frac{21}{24}=\frac{7}{8}$. *Ans.*

EXAMPLE. — Reduce $\frac{40}{2}$ two third ounces to an equivalent fraction in pounds.

$$\frac{40}{2} \times \frac{2}{3} = \frac{80}{6} \div \frac{16}{1} = \frac{80}{96} = \frac{5}{6}. \quad \textit{Ans.}$$

$$\text{Or, } \frac{40}{2} \times \frac{2}{3} \times \frac{1}{16} = \frac{80}{96} = \frac{5}{6}. \quad \textit{Ans.}$$

To reduce a fraction in a higher to whole numbers in lower denominations.

RULE. — Multiply the numerator of the given fraction by the number of parts of the next lower denomination that make ONE of the given fraction, and divide the product by the denominator. Multiply the numerator of the fractional part of the quotient thus obtained by the number of parts in the next lower denomination that make 1 of the denomination of the quotient, and divide by its denominator for whole numbers as before; so proceed until the whole numbers in each denomination desired are obtained.

EXAMPLE. — How many hours, minutes, and seconds, in $\frac{9}{14}$ of a day?

$$\frac{9 \times 24}{14} = \frac{216}{14} = 15, \ \frac{3 \times 60}{7} = \frac{180}{7} = 25, \ \frac{5 \times 60}{7} = \frac{300}{7} = 42\tfrac{6}{7}, =$$

15 h., 25 m., $42\frac{6}{7}$ sec. *Ans.*

EXAMPLE. — How many minutes in $\frac{9}{14}$ of a day?

$$\frac{9 \times 24 \times 60}{14} = \frac{12960}{14} = 925\tfrac{5}{7}. \quad \textit{Ans.}$$

To reduce fractions, or whole numbers and fractions, in lower denominations, to their value in a higher denomination.

RULE. — Reduce the mixed numbers to improper fractions, find their common denominator, and change each whole number and numerator to correspond therewith. Then reduce the higher numbers to their values in the lowest denomination, add the value in the lowest denomination thereto, and take their sum for a numerator. Multiply the common denominator by the number required of the lowest denomination to make ONE of the next higher, that product by the number required of that denomination to make 1 of the next higher, and so on, until the highest denomination desired is reached, and take the product for a denominator, and reduce to lowest terms.

EXAMPLE. — Reduce $5\frac{1}{3}$ oz., $3\frac{1}{5}$ dwts., $2\frac{1}{2}$ grs., troy, to lbs.

$\frac{16}{3} \cdot \frac{16}{5} \cdot \frac{5}{2} = \frac{160 \cdot 96 \cdot 75}{30}$; therefore,

$$\begin{array}{r} 160 \times 20 = 3200 \\ 96 \\ \hline 3296 \times 24 = 79104 \\ 75 \\ \hline 79179 \\ \hline 30 \times 24 \times 20 \times 12 = 172800 \end{array} \Bigg\} = .458 + \text{lbs. } \textit{Ans}$$

EXAMPLE. — Reduce 11 hours, 59 minutes, 60 seconds, to the fraction of a day.

$$
\begin{array}{r}
11 \times 60 = 660 \\
59 \\
\hline
719 \times 60 = 43140 \\
60 \\
\hline
43200 \\
60 \times 60 \times 24 = 86400
\end{array}
\Bigg\} = \tfrac{1}{2}. \text{ Ans.}
$$

EXAMPLE. — Reduce 15 h., 25 m., $42\frac{6}{7}$ sec., to the fraction of a day.

$$
\begin{array}{r}
15 \times 60 \times 60 = 54000 \\
25 \times 60 = 1500 \\
42\frac{6}{7} \\
\hline
55542\frac{6}{7} \\
7 \\
\hline
388800 \\
7 \times 60 \times 60 \times 24 = 604800
\end{array}
\Bigg\} = \tfrac{9}{14}. \text{ Ans.}
$$

To work fractions, or whole numbers and fractions, by the Rule of Three, or Proportion.

RULE. — Reduce the mixed terms to simple fractions, state the question as in whole numbers, invert the divisor, and multiply and divide as in whole numbers.

EXAMPLE. — If $2\frac{1}{2}$ yards of cassimere cost \$$4\frac{1}{4}$, what will $\frac{3}{4}$ of a yard cost? $2\frac{1}{2} = \frac{5}{2}$; $4\frac{1}{4} = \frac{17}{4}$; then,

$\frac{5}{2} : \frac{17}{4} :: \frac{3}{4} : x, = \frac{17}{4} \times \frac{3}{4} \times \frac{2}{5} = \frac{102}{80} =$ \$1.27,5. *Ans.*

DECIMAL FRACTIONS.

A decimal fraction is written with its numerator only. Its denominator is understood. It occupies one or more places of figures, and has a point or dot (.) prefixed or placed before it. The dot (.) alone distinguishes it from an integer or whole number. It supposes a denominator whose value is a UNIT broken into parts, having a tenfold relation to the number of places the numerator occupies. The denominator, therefore, of any decimal is always a unit (1) with as many ciphers annexed as the numerator has places of figures. Thus, the denominator of .1, .2, .3, &c., is 10, and the fractions are read, *one tenth*, *two tenths*, *three tenths*, &c. The denominator of .01, .11, .12, &c., is 100, and these are read, *one hundredth*, *eleven hundredths*, *twelve hundredths*, &c. The denominator of .001, .101, .125, &c., is 1000, and these are read *one thousandth, one hundred and one thousandths, one hundred and twenty-five thousandths*, &c. The denominator of a decimal occupying four places of figures as .7525 is 10000, and so on continually.

The first figure on the right of the decimal point is in the place of *tenths*, the second in the place of *tenths* of *tenths*, or *hundredths*, the third in the place of *tenths* of *tenths* of *tenths*, or *thousandths*, &c. Thus the value of a decimal occupying four places of figures, as .7525, for example, is $\frac{7525}{10000}$, $=\frac{752\frac{1}{2}}{1000}$, $=\frac{75\frac{1}{4}}{100}$, $=\frac{7\frac{1}{2}}{10}+\frac{\frac{1}{4}}{100}=\frac{\frac{3}{4}}{1}+\frac{\frac{1}{4}}{100}$. A decimal is converted into a vulgar fraction of equal value, by affixing its denominator.

Ciphers placed on the right of decimals do not change their value. Thus, .1850 = .185, plainly for the reason that the denominator of the latter bears the same relation to that of the former that 185 bears to 1850; from both terms of the fraction a ten fold has been dropped.

Ciphers placed on the left of decimals *decrease* their value ten fold for every cipher so placed. Thus, .1 = $\frac{1}{10}$, .01 = $\frac{1}{100}$, .001 = $\frac{1}{1000}$, &c.

A mixed number is a whole number and a decimal. Thus, 4.25 is a mixed number. Its value is 4 units, or *ones*, and $\frac{25}{100}$ of 1, = $\frac{425}{100}$ = $4\frac{1}{4}$. The number on the left of the separatrix is always a whole number — that on its right, always a decimal.

ADDITION OF DECIMALS.

RULE. — Set the numbers directly under each other according to their values, whole numbers under whole numbers, and decimals under decimals; add as in whole numbers, and point off as many places for decimals in the sum as there are figures in that decimal occupying the greatest number of places.

EXAMPLES. — Add together .125, .34, .1, .8672. Also, 125, 34.11, .235, 1.4322.

```
 .125                 125.
 .34                   34.11
 .1                      .235
 .8672                  1.4322
 ------              ---------
1.4322  Ans.          160.7772  Ans.
```

SUBTRACTION OF DECIMALS.

RULE. — Set the numbers, the less under the greater, and in other respects as directed for addition; subtract as in whole numbers, and point off as many places for decimals in the remainder as the decimal having the greatest number of figures occupies places.

Examples. — Subtract .2653 from .8. Also, 11.5 from 238.134.

.8	238.134
.2653	11.5
.5347 *Ans.*	226.634 *Ans.*

MULTIPLICATION OF DECIMALS.

Rule. — Multiply as in whole numbers, and point off as many places for decimals in the product as there are decimal places in the multiplicand and multiplier both. If the product has not so many places, prefix ciphers to supply the deficiency.

Examples. — Multiply 14.125 by 3.4. Also, 5.14 by .007.

14.125	5.14
3.4	.007
56500	.03598 *Ans.*
42375	
48.0250 = 48.025. *Ans.*	

Note. — Multiplying by a decimal is equivalent to dividing by a whole number that bears the same relation to a UNIT that a unit bears to a decimal. Multiplying by a decimal, therefore, is equivalent to dividing by the denominator of a fraction of equal value whose numerator is 1, or of dividing by the denominator of a fraction of equal value whose numerator is more than 1, and multiplying the quotient by the numerator. Thus, the decimal $.25 = \frac{25}{100} = \frac{1}{4}$, and the decimal $.875 = \frac{875}{1000} = \frac{7}{8}$. And $14.23 \times .25 = 3.5575$, and $14.23 \div 4 = 3.5575$. So, also, $14.23 \times .875 = 12.45125$, and $14.23 \div 8 = 1.77875 \times 7 = 12.45125$. It is sometimes a saving of labor and matter of convenience to achieve multiplication by this process.

DIVISION OF DECIMALS.

Rule. — Write the numbers as for division of whole numbers, then remove the separatrix in the dividend as many places of figures to the right, (supplying the places with ciphers if they are not occupied,) as there are decimal figures in the divisor; consider the divisor a whole number and divide as in division of whole numbers.

Example. — Divide .5 by .17. Also, .129 by 4.

.17).50(2.94+. *Ans.*	4).129(.032+. *Ans.*
34	12
160	9
153	8
70	1
68	
2	

EXAMPLES. — Divide 16.5 by 1.232. Also, 1.2145 by 12.231.

```
1.232,)16.500,(13.3928+. Ans.
       1232
       ----
        4180
        3696
        ----
         4840
         3696
         -----
         11440
         11088
         -----
           3520
           2464
           -----
           10560
            9856
            ----
             704
```

```
12.231,)1.214,50(.09929+ } Ans.
        1 100 79  .0993— }
        --------
          113 710
          110 079
          -------
            3 6310
            2 4462
            -------
            1 18480
            1 10079
            -------
               8401
```

NOTE. — Dividing by a decimal is equivalent to multiplying by a whole number that bears the same proportion to a UNIT that a unit bears to the decimal. Dividing by a decimal, therefore, is equivalent to multiplying by the denominator of a fraction of equal value whose numerator is 1, or multiplying by the denominator of a fraction of equal value whose numerator is more than 1, and dividing the product by the numerator. Dividing by a fraction is equivalent to multiplying by its denominator and dividing the product by its numerator, or dividing by its numerator and multiplying the quotient by its denominator. Thus, $.5 = \frac{5}{10} = \frac{1}{2}$, and $.75 = \frac{75}{100} = \frac{3}{4}$. And $12.24 \div .5 = 24.48$, and $12.24 \times 2 = 24.48$. So, also, $12.24 \div .75 = 16.32$, and $12.24 \times 4 = 48.96 \div 3 = 16.32$. This method of accomplishing division may often be resorted to with convenience.

REDUCTION OF DECIMALS.

To reduce a decimal in a higher to whole numbers in successive lower denominations.

RULE. — Multiply the decimal by that number in the next lower denomination that equals ONE of the denomination of the decimal, and point off as many places for a remainder as the decimal so multiplied has places. Multiply the remainder by the number in the next lower denomination that equals 1 of the denomination of the remainder, and point off as before; so continue, until the reduction is carried to the lowest denomination required.

EXAMPLE. — What is the value of .62525 of a dollar?

```
        .62525
           100
       --------
Cents,  62.52500
              10
       --------
Mills,   5.25000     Ans. 62 cents 5¼ mills.
```

EXAMPLE. — What is the value of .46325 of a barrel?

	.46325
	32
Gallons,	14.82400
	4
Quarts,	3.296
	2
Pints,	.592
	4
Gills,	2.368. *Ans.* 14 gals. 3 qts. $2\frac{368}{1000}$ gills.

EXAMPLE. — How many pence in .875 of a pound?

$.875 \times 240 = 210.$ *Ans.*

To reduce decimals, or whole numbers and decimals, in lower denominations, to their value in a higher denomination.

RULE. — Reduce all the given denominations to their value in the lowest denomination, then divide their sum by the number required of the lowest denomination to make ONE of the denomination to which the whole is to be reduced.

EXAMPLE. — Reduce 14 gallons, 3 quarts, 2.368 gills, to the decimal of a barrel.

$14 \times 4 = 56 + 3 = 59 \times 8 = 472 + 2.368 = 474.368.$

$8 \times 4 \times 32 = 1024$) 474.368 (.46325. *Ans.*

To work decimals, or whole numbers and decimals, by the Rule of Three, or Proportion.

RULE. — State the question and work it as in whole numbers, taking care to point off as many places for decimals in the product to be used as the dividend, as there are decimals in the two terms which form it, and to remove the decimal point therein as many places to the right as there are decimals in the term to be used as a divisor, before the division is had.

EXAMPLE. — If .75 of a pound of copper is worth .31 of a dollar how much is 3.75 lbs. worth?

.75 : .31 :: 3.75
.31
375
1125
.75) 1.16,25 ($1.55. *Ans.*

PROPORTION, OR RULE OF THREE.

THE RULE OF PROPORTION involves the employment of three terms — a divisor and two factors for forming a dividend — and seeks a quotient, which, when the proposition is written in ratio, bears the same relation to the third term that the second term bears to the first. Two of the terms given are of like name or nature, and the other is of the name or nature of the quotient or answer sought. That of the nature of the answer is always one of the factors for forming the dividend, and, if the answer is to be greater than that term, the larger of the remaining two is the other; but if the answer is to be less than that term, the less of the remaining two is the other — the remaining term is the divisor.

EXAMPLE. — If $12 buy 4 yards of cloth, how many yards will $108 buy?

$$\frac{\not{4} \times 108}{\not{12}_{3}} = \frac{108}{3} = 36 \text{ yards. } Ans.$$

EXAMPLE. — If 4 yards of cloth cost $12, how many dollars will 36 yards cost?

$$\frac{12 \times 36}{4} = 108 \text{ dollars. } Ans.$$

EXAMPLE. — If 30 men can finish a piece of work in 12 days, how many men will be required to finish it in 8 days?

$$\frac{30 \times 12}{8} = 45 \text{ men. } Ans.$$

EXAMPLE. — If 45 men require 8 days to finish a piece of work, how many men will finish the same work in 12 days?

$$\frac{45 \times 8}{12} = 30 \text{ men. } Ans.$$

EXAMPLE. — If 8 days are required by 45 men to finish a piece of work, how many days will be required by 30 men to finish the same work?

$$\frac{8 \times 45}{30} = 12 \text{ days. } Ans.$$

EXAMPLE. — If 12 days are required by 30 men to perform a piece of work, how many days will be required by 45 men to do the same work?

$$\frac{12 \times 30}{45} = 8 \text{ days. } Ans.$$

EXAMPLE. — I borrowed of my friend $150, which I kept 3 months, and, on returning it, lent him $200; how long may he keep the sum that the interest, at the same rate per cent., may amount to that which his own would have drawn?

$$150 \times 3 \div 200 = 2\tfrac{1}{4} \text{ months.} \quad \textit{Ans.}$$

EXAMPLE. — A garrison of 250 men is provided with provisions for 30 days, how many men must be sent out that the provisions may last those remaining 42 days?

$$250 \times 30 \div 42 = 179, \text{ and } 250 - 179 = 71. \quad \textit{Ans.}$$

EXAMPLE. — If to the short arm of a lever 2 inches from the fulcrum there be suspended a weight of 100 lbs., what power on the long arm of the lever 20 inches from the fulcrum will be required to raise it?

$$20 : 2 :: 100 = 10 \text{ lbs.} \quad \textit{Ans.}$$

EXAMPLE. — At what distance from the fulcrum on the long arm of a lever must I place a pound weight, to equipoise or weigh 20 lbs. suspended 2 inches from the fulcrum at the other end?

$$1 : 2 :: 20 : 40 \text{ inches.} \quad \textit{Ans.}$$

NOTE. — If we examine the foregoing with reference to the fact, we shall see that every proposition in simple proportion consists of a *term and a half!* or, in other words, of a *compound* term consisting of two factors, and a factor for which another factor is sought that together shall equal the compound. We have only to multiply the factors of the compound together — and a little observation will enable us to distinguish it — and divide by the remaining factor, and the work is accomplished. See COMPOUND PROPORTION.

COMPOUND PROPORTION, OR DOUBLE RULE OF THREE.

COMPOUND PROPORTION, like *single* proportion, consists of THREE terms given by which to find a fourth — a divisor and two factors for forming a dividend — but unlike single proportion, one or more of the terms is a compound, or consists of two or more factors; and sometimes a portion of the fourth term is given, which, however, is always a part of the divisor.

Of the given terms, two are suppositive, dissimilar in their natures, and relate to each other, and to each other only; and upon their relation the whole is made to depend; the remaining term is of the nature of one of the former, and relates to the fourth term, which is of the nature of the other.

The object sought is a number, which, multiplied into the factor or factors of the fourth term given, if any, and if not, which of itself, bears the same proportion to the dissimilar term to which it relates, as the suppositive term of like nature bears to the term to which it relates.

RULE. — Observe the denomination in which the demand is made, and of the suppositive terms make that of like nature the second, and the other the first; make the remaining term the third term; and, if there are any factors pertaining to the fourth term, affix them to the first; multiply the second and third terms together and divide by the first, and the quotient is the answer, term, or portion of a term, sought.

EXAMPLE.—If 12 horses in 6 days consume 36 bushels of oats how many bushels will suffice 21 horses 7 days?

$$12 \times 6 : 36 :: 21 \times 7 : x.$$

$$\frac{\overset{3}{\cancel{36}} \times 21 \times 7}{\cancel{12} \times \underset{2}{\cancel{6}}} = \frac{147}{2} = 73\tfrac{1}{2} \text{ bushels. } \textit{Ans.}$$

EXAMPLE.—If 12 horses in 6 days consume 36 bushels of oats, how many horses will consume $73\frac{1}{2}$ bushels in 7 days?

$$36 : 12 \times 6 :: 73\tfrac{1}{2} : 7 \times x.$$

$$\frac{12 \times 6 \times 73\frac{1}{2}}{36 \times 7} = \frac{147}{7} = 21 \text{ horses. } \textit{Ans.}$$

EXAMPLE.—If the interest on \$1 is 1.4 cts. for 73 days, (exact interest at 7 per cent.,) what will be the interest on \$150.42 for 146 days?

$$73 : 1.4 :: 150.42 \times 146 : x.$$

$$\frac{1.4 \times 150.42 \times 146}{73} = \$4.21. \ \textit{Ans.}$$

EXAMPLE.—If the interest on \$1 is 1.2 cts. for 73 days, (exact interest at 6 per cent.,) what will be the interest on \$125 for 90 days?

$$73 : 1.2 :: 125 \times 90 : x = \$1.85. \ \textit{Ans.}$$

EXAMPLE.—If \$100 at 7 per cent. gain \$1.75 in 3 months, how much at 6 per cent. will \$170 gain in $11\frac{1}{2}$ months?

$$100 \times 7 \times 3 : 1.75 :: 170 \times 6 \times 11.5 : x.$$

$$1.75 \times 170 \times 6 \times 11.5 \div 100 \times 7 \times 3 = \$9.77,5. \ \textit{Ans.}$$

EXAMPLE.—By working 10 hours a day 6 men laid 22 rods of wall in 3 days; how many men at that rate, who work but 9 hours a day, will lay 40 rods of wall in 8 days?

$$22 : 6 \times 3 \times 10 :: 40 : 9 \times 8 \times x.$$

$$6 \times 3 \times 10 \times 40 \div 22 \times 9 \times 8 = 4\tfrac{6}{11}. \ \textit{Ans.}$$

EXAMPLE.—If it costs \$112 to keep 16 horses 30 days, and it costs as much to keep 2 horses as it costs to keep 5 oxen, how much will it cost to keep 28 oxen 36 days?

$$16 \times 30 : 112 :: \tfrac{2}{5} \times 28 \times 26 : x.$$

$$\text{Or,} — 16 \times 30 \times 5 : 112 :: 28 \times 36 \times 2 : x.$$

$$\frac{\overset{7}{\not{112}}\ \ 28\ \ \overset{12}{\not{36}}\ \ \not{2}}{\not{16}\ \ \not{30}\ \ 5} = \frac{28 \times 12 \times 7}{5 \times 5} = \$94.08. \quad \textit{Ans}$$

$$\not{15}$$

$$5$$

EXAMPLE. — If 24 men, in 8 days of 10 hours each, can dig a trench 250 feet long, 8 feet wide, and 4 feet deep, how many men, in 12 days of eight hours each, will be required to dig a trench 80 feet long, 6 feet wide, and 4 feet deep?

$$250 \times 8 \times 4 : 24 \times 8 \times 10 :: 80 \times 6 \times 4 : 12 \times 8 \times x = 5—. \quad \textit{Ans.}$$

EXAMPLE. — If 120 men in six months perform a given task, working 10 hours a day, how many men will be required to accomplish a like task in 5 months, working 9 hours a day?

$$120 \times 6 \times 10 = 5 \times 9 \times x.$$

$$\text{Or,} — 1 : 120 \times 6 \times 10 :: 1 : 5 \times 9 \times x. = 160. \quad \textit{Ans.}$$

EXAMPLE. — The weight of a bar of wrought iron, 1 foot in length, 1 inch in breadth, and 1 inch thick, being 3.38 lbs., (and it is so,) what will be the weight of that bar whose length is $12\frac{1}{2}$ feet, breadth $3\frac{1}{4}$ inches, and thickness $\frac{3}{4}$ of an inch?

$$1 : 3.38 :: 12.5 \times 3.25 \times .75 : x.$$

$$\text{Or,} — 1 : 3.38 :: \tfrac{25}{2} \times \tfrac{13}{4} \times \tfrac{3}{4} : x, \text{ and}$$

$$\frac{3.38 \times 25 \times 13 \times 3}{2 \times 4 \times 4} = 102.98+ \text{ lbs.} \quad \textit{Ans.}$$

EXAMPLE. — The weight of a bar of wrought iron, one foot in length and 1 inch square, being 3.38 lbs., what length shall I cut from a bar whose breadth is $2\frac{3}{4}$ inches, and thickness $\frac{1}{2}$ inch, in order to obtain 10 lbs.? $3.38 : 1 :: 10 : \frac{11}{4} \times \frac{1}{2} \times x.$

$$\frac{1 \times 10 \times 4 \times 2}{3.38 \times 11 \times 1} = 2 \text{ feet } 1\tfrac{8}{10} \text{ inches.} \quad \textit{Ans.}$$

CONJOINED PROPORTION, OR CHAIN RULE.

THE CHAIN RULE is a process for determining the value of a given quantity in one denomination of value, in some other given denomination of value; or the immediate relationship which exists between two denominations of value, by means of a *chain* of approximate steps, circumstances, or equivalent values, known to exist, which connect them. In every instance at least *five* terms or values are employed in the process, and in all instances the number employed will be uneven. A proposition involving but three terms, of this nature, is a question in single proportion. The equivalent values employed are divided into *antecedents* and *consequents*, or causes and effects; and the value or quantity for which an equivalent is sought, is called the odd term.

RULE. — 1. *When the value in the denomination of the first antecedent is sought of a given quantity in the denomination of the last consequent.* — Multiply all the antecedents and the odd term together for a dividend, and all the consequents together for a divisor; the quotient will be the answer or equivalent sought.

RULE. — 2. *When the value in the denomination of the last consequent is sought of a given quantity in the denomination of the first antecedent.* — Multiply all the consequents and the odd term together for a dividend, and all the antecedents together for a divisor; the quotient will be the answer required.

EXAMPLE. — I am required to give the value, in Federal money, of 5 Canada shillings, and know no immediate connection or relationship between the two currencies — that of Canada and that of the United States. The nearest that I do know is that 20 Canada shillings have a value equal to 32 New York shillings, and that 12 New York shillings equal in value 9 New England shillings, and that 15 New England shillings equal \$2.50; and with this knowledge will seek the value, in Federal money, of the 5 Canada shillings.

$$\frac{2.50 \times 9 \times 32 \times 5}{15 \times 12 \times 20} = \$1. \quad \textit{Ans.}$$

EXAMPLE. — If \$2½ equal 15 New England shillings, and nine shillings in New England equal 12 shillings in New York, and 32 shillings in New York equal 20 shillings in Canada, how many shillings in Canada will equal \$1?

$$\begin{array}{cccc} & \not{3} & \not{8} & \\ 15 & \not{1}\not{2} & \not{2}\not{0} & 1 \\ \hline \not{2}\frac{1}{2} & \not{9} & \not{3}\not{2} & \\ & 3 & \not{4} & \end{array} = \frac{15}{3} = 5 \text{ shillings.} \quad \textit{Ans.}$$

EXAMPLE. — If 14 bushels of wheat weigh as much as 15 bushels of fine salt, and 10 bushels of fine salt as much as 7 bushels of coarse, and 7 bushels of coarse salt as much as 4 bushels of sand, how many bushels of sand will weigh as much as 40 bushels of wheat?

$$\frac{15 \times 7 \times 4 \times 40}{14 \times 10 \times 7} = 17\frac{1}{7} \text{ bushels.} \quad \textit{Ans.}$$

PERCENTAGE.

Pure percentage, or PERCENTAGE, is a rate by the hundred of a *part* of a quantity or number denominated the principal, or basis. But percentage, considered as a means, and as commonly applied, is mixed and related in an eminent degree; and in this light may be regarded as divided into orders bearing different names.

Thus *Interest* is percentage related to intervals of time in the past.

Discount is percentage related to interest, and intervals of time in the future.

Profit and Loss is comparative percentage, or percentage related to the positive and negative interests in business, etc., etc.

Pure percentage is commonly called BROKERAGE when paid to a broker for services in his line.

It is called COMMISSION when paid to or received by a factor or commission merchant for buying or selling goods.

It is called PREMIUM by an insurance company, when taken for insuring against loss.

It is called PRIMAGE when it is a charge in addition to the freight of a vessel, etc.

Comparative percentage relates to the differences of quantities, and is confined always to the idea of *more* or *less*. It implies ratio This description of percentage, though much in practice, seems not to be well understood; and often a quantity is indirectly stated to be many times less than nothing, or many times greater than it is. The difference of two quantities cannot be as great as a hundred per cent. of the greater, however widely unequal the quantities may be, nor as small as no per cent. of the greater or lesser, however nearly equal they may be. No quantity or number can be as small as 1 time less than another quantity or number; and therefore cannot be as small as 100 per cent. less. But, since one quantity may be many by 1 time, or many times greater than another with which it is compared, it may be said to be many by 100 times, or many hundred per cent. greater.

When one of two quantities in comparison is stated to be three times less, or three hnndred per cent. less, for instance, than the other, the expression is incorrect and absurd. The meaning evidently is, that it is two-thirds less, or only one-third as large as the other, — that it is $66\frac{2}{3}$ per cent. less, or only $33\frac{1}{3}$ per cent. as large as the other. In common comparison, 1 is the measuring unit. In percentage, 100 is the measuring unit.

Let $a =$ principal.
$b =$ percentage.
$s =$ amount (sum of the principal and percentage).
$d =$ difference of the principal and percentage.
$r =$ rate of the percentage.
$p =$ rate per cent. of the percentage.

$$a = s - b = b \div r = 100b \div p = 100s \div (100 + p),$$
$$b = s - a = ar = ap \div 100,$$
$$p = 100r = 100b \div a = 100(s - a) \div a,$$
$$r = p \div 100 = b \div a = (s - a) \div a,$$
$$s = a + b = a(1 + r) = a(100 + p) \div 100,$$
$$d = a - b = 2a - s = s - 2b = a(1 - r).$$

To find the Percentage.

EXAMPLES.

What is $\frac{1}{4}$ of 1 per cent. of \$200?
$b = ar = ap \div 100 = \$0.50$. *Ans.*

$\frac{8}{7}$ of 2 per cent. of 50 is what part of 50?

$$\frac{50 \times 8 \times 2}{7 \times 100} = 1\tfrac{1}{7}. \quad \textit{Ans.}$$

What is $\frac{3}{5}$ of $\frac{5}{3}$ of $\frac{1}{2}$ of 24 per cent. of 150 lbs.?
$150 \times 12 \div 100 = 18$ lbs. *Ans.*

What is $2\frac{3}{8}$ per cent. of 19 bushels?
$\frac{19}{8} \times \frac{19}{100} = 0.45125$ bushels. *Ans.*

Bought a job lot of merchandise for \$850, and sold it the same day, brokerage, $2\frac{1}{2}$ per cent., for \$975; what was the net gain?

$$s - sr - a = s - (sr + a) = s(1 - r) - a = 975 - 975 \times .025 - 850 = \$100.625. \quad \textit{Ans.}$$

To find the Rate or Rate Per Cent.

EXAMPLES.

What per cent. of \$20 is \$2?
$r = b \div a,\ p = 100b \div a = 10$ per cent. *Ans.*

12 dozen is equal to what per cent. of 2 dozen?
$12 \div 2 = 6$, 600 per cent. *Ans.*

What part of $5\frac{1}{2}$ lbs. is $\frac{3}{4}$ of 2 lbs. ?

$$r = \tfrac{3}{4} \times \tfrac{2}{11} = \tfrac{1.5}{5.5} = 0.27\tfrac{3}{11}. \quad \textit{Ans.}$$

$24\frac{1}{2}$ per cent. is what per cent. of $36\frac{3}{4}$ per cent. ?

$66\frac{2}{3}$ per cent. *Ans.*

For an article that cost \$4, \$5 were received; what per cent of \$4 was received ?

$$p = 5 \times 100 \div 4 = 125 \text{ per cent.} \quad \textit{Ans.}$$

A farmer sowed 4 bushels of wheat, which produced 48 bushels; what per cent. was the *increase?* 48 is *more* than 4 by what per cent. of 4 ? The difference of 48 and 4 is what per cent. of 4 ?

$$r = \frac{a-b}{b} = \frac{a}{b} - 1,\ p = \frac{100(a-b)}{p} = \frac{48-4}{4} = 48 \div 4 - 1 =$$

$$100(48-4) \div 4 = 1100 \text{ per cent.} \quad \textit{Ans.}$$

What per cent. would have been the *decrease*, if he had sowed 48 bushels, and harvested only 4 bushels ? 4 is *less* than 48 by what rate of 48 ? The difference of 48 and 4 is what per cent. of 48 ?

$$r = (a-b) \div a = 1 - \frac{b}{a} = 0.91\tfrac{2}{3}, \text{ or } 91\tfrac{2}{3} \text{ per cent.} \quad \textit{Ans.}$$

Since water is composed of 8 atoms of oxygen and 1 atom of hydrogen, what per cent. of it is oxygen ? 8 is what per cent. of the sum of 8 and 1 ?

$$r = \frac{a}{a+b} = 1 - \frac{b}{a+b},\ p = \frac{100a}{a+b} = \frac{8}{8+1} = .8889 -,$$

or 88.89 – per cent. *Ans.*

What per cent. of it is hydrogen ? 1 is what per cent. of the sum of 8 and 1 ?

$$r = 1 - \frac{a}{a+b} = \frac{b}{a+b},\ p = \frac{100b}{a+b} = \frac{1}{8+1} = .1111 +, \text{ or}$$

11.11 + per cent. *Ans.*

How many volumes of water must be added to 100 volumes of 90 per cent. alcohol to reduce it to 50 per cent. alcohol or common proof? 90 is more than 50 by what per cent. of 50 ? The difference of 90 and 50 is what per cent. of 50?

$$p = \frac{(a-b)100}{b} = \frac{(90-50)100}{50} = 80. \quad \textit{Ans.}$$

How many volumes of 50 per cent. alcohol must be added to 100 volumes of 90 per cent. alcohol to produce 80 per cent. alcohol? 90 is more than 80 by what per cent. of the difference of 80 and 50? The difference of 90 and 80 is what per cent. of the difference of 80 and 50?

$$p = \frac{(a-b)100}{b-b'} = \frac{(90-80)100}{80-50} = 33\tfrac{1}{3}. \quad \textit{Ans.}$$

How many volumes of 90 per cent. alcohol must be added to 100 volumes of 50 per cent. alcohol to raise it to 80 per cent. alcohol? 50 is less than 80 by what per cent. of the difference of 90 and 80? The difference of 80 and 50 is what per cent. of the difference of 90 and 80?

$$\frac{(b-b')100}{a-b} = \frac{(80-50)100}{90-80} = 300. \quad \textit{Ans.}$$

If to 2 volumes of 95 per cent. alcohol, 1 volume of 50 per cent. alcohol be added, what per cent. alcohol will be the mixture? The sum of 50 and twice 95 is what per cent. of the sum of 2 and 1?

$$\frac{2a+b}{2+1} = \frac{2 \times 95 + 50}{2+1} = 80 \text{ per cent.} \quad \textit{Ans.}$$

In a barrel of apples, the number of sound ones was 60 per cent. *greater* than the number that were damaged. What per cent. *less* was the number that were damaged than the number that were sound? 60 per cent. is what per cent. of the sum of 100 per cent. and 60 per cent.? .6 is what rate of 1 + .6?

$$r = \frac{a}{1+a} = 1 - \frac{100}{1+a} = \frac{0.a}{1+.a} = 1 - \frac{1}{1.a} = \frac{60}{1+60} = .375, \text{ or}$$

$37\frac{1}{2}$ per cent. *Ans.*

Since the number of damaged apples was $37\frac{1}{2}$ per cent. less than the number that were sound, what per cent. greater was the number that were sound than the number that were damaged?

$$r = a \div (1-a) = 1 \div (1-a) - 1 = 60 \text{ per cent.} \quad \textit{Ans.}$$

Since the number of sound ones was 60 per cent. greater than the number that were damaged, what per cent. of the whole were sound?

$$r = \frac{a+a^2}{2a} = \frac{1+.a}{2}, \; p = \frac{100+60}{2} = 80 \text{ per cent.} \quad \textit{Ans.}$$

What per cent. of the whole were damaged?

$$(100-60) \div 2 = 20 \text{ per cent.} \quad \textit{Ans.}$$

Since 20 per cent. of the apples were damaged, what per cent. less was the number that were damaged than the number that were sound?

$$r = \frac{1 - 2.a}{2 - 2.a} = 1 - \frac{1}{2 - 2.a},\ p = \frac{100 - 2a}{200 - 2a} = 100 - \frac{100}{200 - 40} =$$

$37\frac{1}{2}$ per cent. *Ans.*

What per cent. greater was the number that were sound than the number that were damaged?

$$r = 2 - (1 + 2.a) = 2 - 2.a - 1 = 60 \text{ per cent.} \quad \textit{Ans.}$$

Since 80 per cent. of the whole were sound, what per cent. less was the number that were damaged than the number that were sound?

$$r = \frac{2.a - 1}{2.a} = 1 - \frac{1}{2.a} = \frac{2 \times .80 - 1}{2 \times .80} = 37\tfrac{1}{2} \text{ per cent.} \quad \textit{Ans.}$$

Since the number of damaged ones was $37\frac{1}{2}$ per cent. less than the number that were sound, what per cent. of the whole were sound?

$$r = \frac{1}{2 - 2.a},\ p = \frac{100}{2 - 2a} = \frac{100}{2 - 2 \times 37.5} = 80 \text{ per cent.} \quad \textit{Ans.}$$

Since 80 per cent. of the whole were sound, what per cent. greater was the number that were sound than the number that were damaged?

$$r = \frac{2 - .a}{2} = 2.a - 1 = 2 \times .80 - 1 = 60 \text{ per cent.} \quad \textit{Ans.}$$

Lost 20 per cent. of a cargo of coal by jettison, and 5 per cent. of the remainder by screening, what per cent. of the coal was saved?

$$\left.\begin{matrix} a - b' = d' \\ d' - b'' = d'' \end{matrix}\right\} \begin{matrix} r = (1 - r')\,(1 - r'') = (1 - .20) - (1 - .20) \\ \times .05 = (1 - .20)(1 - .05) = 76 \text{ per cent. } \textit{Ans.} \end{matrix}$$

$d'' - b''' = d'''$, &c.

Yesterday drew 12 per cent. of my balance of $4,273 in the bank, and deposited $1,000; and to-day have drawn $31\frac{1}{4}$ per cent. of the balance left over, or as it stood last night. What per cent. of the sum of the first-mentioned balance and deposit of yesterday have I drawn?

$$r = \frac{b' + b''}{a + m} = \frac{512 + 1487.575}{4273 + 1000} = 37.9354 + \text{ per cent.} \quad \textit{Ans.}$$

What per cent. of the said sum is remaining in the bank ?

$$1-\frac{b+b''}{a+m}=\frac{a+m-b'-b''}{a+m}=\frac{a+m-(b'+b'')}{a+m}=62.0646-$$ per cent. *Ans.*

What per cent., predicating it upon the first-mentioned balance, have I drawn ?

$$r=\frac{b'+b''}{a}=\frac{512.76+1487.576}{4273}=46.8134-$$ per cent. *Ans.*

What per cent. have I drawn, predicating it upon what I now have in the bank ?

$$r=\frac{b'+b''}{a-b'+m-b''}=\frac{b'+b''}{a+m-(b'+b'')}=61.1225+$$ per cent. *Ans.*

What amount of money must I deposit to make good $62\frac{1}{2}$ per cent. of the aforementioned sum ?

$$d=r(a+m)+b'+b''-(a+m)=r(a+m)-d''=$$ \$22.96. *Ans.*

To find the Principal or Basis.

EXAMPLES.

The percentage being 250, and the rate .06, what is the principal ?

$a=b\div r=100b\div p=250\div .06=25{,}000\div 6=4{,}166\frac{2}{3}$. *Ans.*

A tax at the rate of $\frac{5}{6}$ of 1 per cent. on the valuation was \$27.50. What was the valuation ?

$$a=\frac{b\times 6\times 100}{5}=\$3{,}300.$$ *Ans.*

Sold 120 barrels of flour, which amounted to 12 per cent. of a certain consignment. The consignment consisted of how many barrels ?

$120\div 0.12=1{,}000$. *Ans.*

216 bushels is *more* by 8 per cent., or 8 per cent. more than what number of bushels ? 8 per cent. more than what number is equal to 216 ? What number, plus 8 per cent. of it, will make 216 ?

$$a=s\div(1+r)=216\div 1.08=200.$$ *Ans.*

200 lbs. is *less* by 8 per cent., or 8 per cent. less, than what num-

ber of lbs.? 8 per cent. less than what number is 200? What number, minus 8 per cent. of it, is equal to 200?

$$a = d \div (1 - r) = 200 \div (1 - .08) = 217\tfrac{9}{23}. \quad \textit{Ans.}$$

$$\therefore 217\tfrac{9}{23} - 217\tfrac{9}{23} \times .08 = 200 = a - b = d = a\,(1 - r).$$

To a quantity of silver, a quantity of copper equal to 20 per cent. of the silver is to be added, and the mass is to weigh 22 ounces. What weight of silver is required?

$$a = s \div (1 + r) = 22 \div 1.2 = 18\tfrac{1}{3} \text{ ounces.} \quad \textit{Ans.}$$

What weight of copper is required?

$$s - \frac{s}{1+r} = \frac{sr}{1+r} = 3\tfrac{2}{3} \text{ ounces.} \quad \textit{Ans.}$$

To a quantity of copper, a quantity of nickel equal to $62\frac{1}{2}$ per cent. of the copper, a quantity of zinc equal to $33\frac{1}{3}$ per cent. of the copper, and a quantity of lead equal to 5 per cent. of the copper, are to be added; and the whole is to weigh $40\frac{1}{6}$ pounds. The weight of each constituent of the alloy is required.

$$a = \frac{s}{1 + r + r' + r''} = \frac{40\frac{1}{6}}{1 + .62\frac{1}{2} + .33\frac{1}{3} + .05}$$

$$\left.\begin{aligned} &= 20 \text{ lbs. of copper,} \\ b &= 20\,r = 12\tfrac{1}{2} \text{ lbs. of nickel,} \\ b' &= 20\,r' = 6\tfrac{2}{3} \text{ lbs. of zinc,} \\ b'' &= 20\,r'' = 1 \text{ lb. of lead.} \end{aligned}\right\} \textit{Ans.}$$

INTEREST.

Universal for any rate per cent.

T = time in months and decimal parts of a month; t = time in days; P = principal; r = rate per cent., expressed decimally; i = interest.

$$i = \frac{P \times T \times r}{12} = \frac{P \times t \times r}{365}.$$

$$P = \frac{12\,i}{Tr} = \frac{365\,i}{t\,r}. \quad T = \frac{12\,i}{P\,r}. \quad t = \frac{365\,i}{P\,r}. \quad r = \frac{12\,i}{PT} = \frac{365\,i}{Pt}.$$

EXAMPLE. — A promissory note, made April 27, 1864, for

$825\frac{25}{100}$ and interest at 6 per cent., matured Oct. 6, 1865: what was the interest?

Oct. is 10th month. April is 4th month.			
	Y.	m.	d.
	1865 .	10 .	6
	'64 .	4 .	27
Time =	1 .	5 .	9

Time from April 27 to Oct. 6 (one of the dates always included) = 162 days, which, added to the 365 days in the year preceding = 527 days.

NOTE.—One day's interest at least is generally lost by computing the time in years and months, or months, instead of days.

$$825.25 \times 17.3 \times .06 \div 12 = \$71.38. \quad \textit{Ans.}$$

$$825.25 \times 527 \times .06 \div 365 = \$71.49. \quad \textit{Ans.}$$

To find a constant divisor, k, for any given rate per cent.

When the time is taken in months, $k = 12 \div r$.
When the time is taken in days, $k = 365 \div r$; thus,

When the RATE *is* 6 *per cent.* $\frac{P \times t}{6083}$ = Interest.

When the RATE *is* 7 *per cent.* $\frac{P \times t}{5214}$ = Interest, &c.

EXAMPLE.—Required the interest on $750 for 93 days, at 7 per cent.

$$750 \times 93 \div 5214 = \$13.38. \quad \textit{Ans.}$$

EXAMPLE.—What is the rate per cent. when $450 gains $94½ in 3 years?

$$450 : 100 :: 94.5 : 3x = 7 \text{ per cent.} \quad \textit{Ans.}$$

$$94.5 \div 3 \times 450 = .07. \quad \textit{Ans.}$$

EXAMPLE.—In what time will $125 at 6 per cent. gain $18¾?

$$6 : 100 :: 18.75 : 125 \times x = 2\tfrac{1}{2} \text{ years.} \quad \textit{Ans.}$$

$$18.75 \div 125 \times .06 = 2\tfrac{1}{2} \text{ years.} \quad \textit{Ans.}$$

EXAMPLE.—What principal at 5 per cent. interest will gain $16⅞ in 18 months?

$$5 : 100 :: 16.875 : 1.5 \times x = \$225. \quad \textit{Ans.}$$

$$16.875 \times 12 \div 18 \times .05 = \$225. \quad \textit{Ans.}$$

When partial payments have been made.

RULE. — Find the amount (sum of the principal and interest) up to the time of the first payment, and deduct the payment therefrom; then find the interest on the remainder up to the next payment, add it to the remainder, or new principal, and from the sum subtract the next payment; and so on for all the payments; then find the amount up to the time of final payment for the final amount.

COMPOUND INTEREST.

If we calculate the interest on a debt for one year, and then on the same debt for another year, and again on the same debt for still another year, the sum will be the *simple* interest on the debt for three years. But, on the contrary, if we calculate the interest on the debt for one year, and then on the *amount* (sum of the principal and interest) for the next year, and then on the second amount for the third year, the sum of the interest so calculated will be the *compound* interest, or yearly compound interest, on the debt for three years; equal to the simple interest on the debt for three years, plus the yearly compound interest on the first year's interest for two years, plus the simple interest on the second year's interest for one year. So, if we divide the time into shorter periods than a year, and proceed for the interest as last suggested, the interest will be compound. Thus we have half-yearly compound interest, or compound interest semi-annually, quarter-yearly compound interest, or compound interest quarterly, &c.

This method of computing interest is predicated upon the natural idea, that interest, when it becomes due by stipulation and is withheld, commences to draw interest, and continues at use to the holder, at the same rate as the principal, until it is paid, like other over-due demands; and that the interest so made matures and becomes due as often, and at the same periods, as that on the principal.

It will be perceived by the foregoing that the *working-time* in compound interest is the interval between the stipulated payments of the interest, or between one stipulated payment of the interest and that of another; and that the *working-rate* is pro rata to the rate per annum.

Thus the *amount* of $100 at semi-annual compound interest for 2 years, at 6 per cent. per annum, is

$$100 \times (1.03)^4 = \$112.550881 = \$112.55, \text{ or}$$

$$\begin{array}{r} 100. \\ .03 \\ \hline 3. \\ 100. \\ \hline 103. \\ .03 \\ \hline 3.09 \\ 103. \\ \hline 106.09 \\ .03 \\ \hline 3.1827 \\ 106.09 \\ \hline 109.2727 \\ .03 \\ \hline 3.278181 \\ 109.2727 \\ \hline \$112.550881, \text{ as before.} \end{array}$$

If we let P = principal or debt at interest,
r = working-rate of interest,
n = number of intervals into which the whole time is divided for the payment of interest, or number of consecutive intervals for the payment of interest that have transpired without a payment having been made,
i = compound interest,
A = P + i or amount, then

$$A = P(1+r)^n; \quad P = \frac{A}{(1+r)^n}; \quad r = \sqrt[n]{\frac{A}{P}} - 1;$$

$$\frac{A}{P} = (1+r)^n; \quad i = A - P.$$

EXAMPLE. — What is the compound interest, or yearly compound interest, on \$100 for $1\frac{1}{2}$ years, at 6 per cent. a year?

$$100 \times 1.06 \times 1.03 = 109.18 - 100 = \$9.18. \quad \textit{Ans.}$$

EXAMPLE. — What is the amount of \$560.46, at 7 per cent. compound interest per year, for 6 years and 57 days?

$$560.46 \times (1.07)^6 \times \left(1 + \frac{.07 \times 57}{365}\right) = \$850.29. \quad \textit{Ans.}$$

EXAMPLE. — The *principal* is $250, the *rate* 8 per cent. a year, the *time* 2 years, and the *interest* compound per quarter year: required the *amount.*

$$250 \times \left(1.\frac{.08}{4}\right)^8 = \$292.91. \quad Ans.$$

When Partial Payments have been made.

RULE. — Find the amount up to the first payment, and deduct the payment therefrom; then find the amount up to the next payment, and therefrom deduct that payment; and so on for all the payments; then find the amount up to the time of final payment, for the final amount.

EXAMPLE. — A note of hand for $500 and interest from date, at 6 per cent. a year, has been paid in part as follows; viz., two years and four months from the date of the note, by an indorsement of $50; and three years from that indorsement, by an indorsement of $150. It is now eight months since the last payment was made, and the demand is to be settled in full: required the amount at the present time, interest being compound per year.

$$500 \times (1.06)^2 \times 1.02 - 50 = 523.036$$

$$\begin{array}{r} 523.036 \\ (1.06)^3 \\ \hline 622.944 \\ 150 \\ \hline 472.944 \\ 1.04 \\ \hline \$491.86. \end{array} \quad Ans.$$

The following table shows $(1+r)$ raised to all the integer powers from 1 to 12 inclusive; r being taken at 4, 5, 6, 7, 8, and 10 per cent. If the numbers in the column headed years are taken to represent years, then 4 per cent., 5 per cent., &c., at the head of the columns of powers, will stand for per cent. per annum: if they are taken to represent half-years, then 4 per cent., 5 per cent., &c., will stand for per cent. per half-year, &c. The quantities in the columns are powers of $(1+r)$, of which the numbers referred to and standing opposite, respectively, are the exponents. Thus, 1.26248, in the 6 per cent. column, and against 4 in the column marked years, $= (1.06)^4$; and so with the others. The powers or quantities in the columns are co-efficients in the calculations.

Years.	4 per cent.	5 per cent.	6 per cent.	7 per cent.	8 per cent.	10 per cent.
1	1.04	1.05	1.06	1.07	1.08	1.10
2	1.0816	1.1025	1.1236	1.1449	1.1664	1.21
3	1.12486	1.15762	1.19102	1.22504	1.25971	1.331
4	1.16986	1.21551	1.26248	1.3108	1.36049	1.4641
5	1.21665	1.27628	1.33823	1.40255	1.46933	1.61051
6	1.26532	1.3401	1.41852	1.50073	1.58687	1.77156
7	1.31593	1.4071	1.50363	1.60578	1.71382	1.94872
8	1.36857	1.47746	1.59385	1.71819	1.85093	2.14359
9	1.42331	1.55133	1.68948	1.83846	1.999	2.35795
10	1.48024	1.62889	1.79085	1.96715	2.15892	2.59374
11	1.53945	1.71034	1.8983	2.10485	2.33164	2.85312
12	1.60103	1.79586	2.0122	2.25219	2.51817	3.13843

NOTE. — If a co-efficient is wanted for a greater number of years or intervals of time than is given in the table, square the tabular co-efficient opposite half that number of intervals, or cube the tabular co-efficient opposite one-third that number of intervals, &c., for the co-efficient required. Thus,

$$1.999^2 = 1.58687^3 = 1.08^{12} \times 1.08^6 = 1.08^{18} = 3.996,$$

the co-efficient for 18 years or intervals at 8 per cent. per interval, &c.

If the compound interest alone is sought on a given principal, subtract 1 from the tabular power corresponding to the time and rate, and multiply the remainder by the given principal; the product will be the compound interest. Thus $(1.26532 - 1) \times 100 = \$26.532$, the yearly compound interest, at 4 per cent. per annum, on \$100 for 6 years, or the half-yearly compound interest, at 8 per cent. per annum, on \$100 for 3 years, or the half-yearly compound interest, at 4 per cent. per half year, on \$100 for 6 half-years.

EXAMPLE. — What is the amount of \$125.54, at 5 per cent. compound interest, for 7 years, 21 days?

$1 + \frac{21 \times .05}{365} = 1.00288$, the co-efficient for the odd days; and, turning to the 5 per cent. column in the table, we find against 7, in the column of years, 1.4071, the co-efficient for 7 years: then

$$125.54 \times 1.4071 \times 1.00288 = \$178.20. \quad \textit{Ans.}$$

EXAMPLE. — In what time, at 7 per cent. compound interest per annum, will \$1000 gain \$462? $A \div P = (1 + r)^n$: then $1462 \div 1000 = 1.462$, the co-efficient demanded. Turning now to the 7 per cent. column in the table, we find the nearest less co-efficient there (there being none that exactly corresponds) to be that for 5 years; viz., 1.40255. And $\left(\frac{1.462}{1.40255} - 1\right) \div .07 =$ .60553, the fraction of a year over 5 years to the answer.

$$.60553 \times 365 = 221 \text{ days}: \ 5 \text{ years}, 221 \text{ days}. \quad \textit{Ans.}$$

The following TABLE is of the same nature as the preceding, and is applicable when the interest becomes due at regular intervals short of a year, or when the working-rate in compound interest is less than 4 per cent.

The quantities in the $1\frac{3}{4}$ per cent. column apply to quarter-yearly compound interest when the rate is 7 per cent. a year; and those in the $1\frac{1}{4}$ per cent. column, to quarterly compound interest when the rate is 5 per cent. a year; also the former are applicable to monthly compound interest at 21 per cent. per annum, and the latter to monthly compound interest at 15 per cent. per annum; and so relatively, throughout the table.

Times.	3½ per cent.	3 per cent.	2½ per cent.	2 per cent.	1¾ per cent.	1½ per cent.	1¼ per cent.	1 per cent.	½ per cent.
1	1.035	1.03	1.025	1.02	1.0175	1.015	1.0125	1.01	1.005
2	1 07123	1.0609	1.05063	1.0404	1.03531	1.03023	1.02516	1.0201	1.01003
3	1.10872	1.09273	1.07689	1.06121	1.05342	1.04568	1.03797	1.0303	1.01508
4	1.14752	1.12551	1.10381	1.08243	1.07186	1.06136	1.05095	1.0406	1.02015
5	1.18769	1.15927	1.13141	1.10408	1.09062	1.07728	1.06408	1.05101	1.02525
6	1.22925	1.19405	1.15969	1.12616	1.1077	1.09344	1.0774	1.06152	1.03038
7	1.27228	1.22987	1.18869	1.14869	1.12709	1.10984	1.09087	1.07214	1.03553
8	1.31681	1.26677	1.2184	1.17166	1.14681	1.12649	1.10451	1.08286	1.04071
9	1.3629	1.30477	1.24886	1.19509	1.16688	1.14339	1.11631	1.09369	1.04591
10	1.4106	1.34392	1.28008	1.21899	1.1873	1.16054	1.13229	1.10462	1.05114
11	1.45997	1.38423	1.31209	1.24337	1.20808	1.17795	1.14645	1.11567	1.0564
12	1.51107	1.42576	1.34489	1.26824	1.22922	1.19562	1.16078	1.12683	1.06168

EXAMPLE. — What is the amount of $750 for 4 years and 40 days, allowing half-yearly compound interest, at 7 per cent. a year?

In this case, the working-rate for the full periods of time is $3\frac{1}{2}$ per cent., and there are 8 such full periods; then, seeking the co-efficient in the $3\frac{1}{2}$ per cent. column, we find against 8, in the column of times, the quantity or co-efficient 1.31681; and $1 + \frac{40 \times .07}{365} =$ 1.00767: therefore

$$750 \times 1.31681 \times 1.00767 = \$995.18. \quad \textit{Ans.}$$

EXAMPLE. — What is the amount of $1000 at compound interest per quarter-year, at $1\frac{1}{2}$ per cent. per quarter-year, for $4\frac{1}{4}$ years?

$$1000 \times 1.12649^2 \times 1.015 = \$1288.01. \quad \textit{Ans.}$$

BANK INTEREST OR BANK DISCOUNT.

A bank loans money on a promissory note made payable without interest at a future period. The operation is called *discounting* the note at bank, and is as follows: The bank takes the note, finds the interest on it for three days more time than by its own tenor it has to run, subtracts it from the principal, and hands the balance, called the *avails* of the note, in its own bills, to the party soliciting the loan, or offering the note for discount, as it is called; whereby the note becomes the property of the bank, and the maker and indorsers are held for its payment when it matures.

The three days mentioned are called *days of grace*, and the note does not become due to the bank until three days after it becomes due by its own tenor. These proceedings are sanctioned by usage, and protected by law.

Bank interest, then, is bank discount, and bank discount is bank interest. But bank discount is not *discount*, nor is it what is called *legal* interest on the money loaned. It is the interest on the money loaned, plus the interest on the interest of the loan, plus the interest on the difference of the sum taken and the interest on the loan for the time of the loan! A kind of interest more onerous, if any description of interest be onerous, than compound interest, rate for rate and time for time, as may be readily perceived.

Let P = principal or face of the note.
r = working-rate of the interest for the time of the loan.
a = avails of the note or sum borrowed.
i = bank interest.
t = time of the loan.

$R : r :: T : t$. R being the rate per cent. per annum, and T one year.

$P = a \div (1 - r)$. $a = P - Pr$. $i = Pr$. $r = (P - a) \div P$.

If we let n represent the time of the note in months,

$r = \frac{Rn}{12} + \frac{3R}{365}$. But it is the practice with many banks to count the days of grace as so many 360ths of a year.

Putting d to represent the time of the note in days,

$$r = \frac{Rd + 3R}{365}, \text{ true time and rate.}$$

With some banks, it is the practice, in calculating interest, to take the time, when it does not exceed 93 days, as so many 360ths of a year.

A note having 3 months to run from Aug. 10, for instance, will fall due Nov. 10–13; but one having 90 days to run from Aug. 10 will fall Nov. 8–11. The time including grace of the former is 3 mo. 3 ds., and that of the latter 3 mo. 2 ds., mean time. Nevertheless, the former embraces 95 days, or one day more than mean time, and the latter but 93 days.

The following table shows $1-r$, mean time, for the intervals of time set down in the left-hand column; R being taken at 4, 5, 6, 7, and 8 per cent. per annum, as set down at the top of the columns.

Time.		4	5	6	7	8
mo.	ds.	per cent.	per cent.	per cent.	per cent.	per cent.
1	3	.996333	.995417	.9945	.993583	.992667
2	3	.993	.99125	.9895	.98775	.986
3	3	.989667	.987083	.9845	.981917	.979333
4	3	.986333	.982917	.9795	.976083	.972667
5	3	.983	.97875	.9745	.97025	.966
6	3	.979667	.974583	.9695	.964417	.959333
7	3	.976333	.970417	.9645	.958583	.952667
8	3	.973	.96625	.9595	.95275	.946
9	3	.969667	.962083	.9545	.946917	.939333
10	3	.966333	.957917	.9495	.941083	.932667
11	3	.963	.95375	.9445	.93525	.926
12	3	.959667	.949583	.9395	.929417	.919333

Putting k to represent the tabular quantity $1-r$,

$$a = Pk,\ P = a \div k,\ i = P - a = P - Pk.$$

EXAMPLE. — What will be the avails of a note for $1,250 payable in 4 months if discounted at a bank, interest being 7 per cent. a year?

The tabular constant $1-r$, in the 7 per cent. column, against 4 months and 3 days in the time column, is .976083, and

$$\$1,250 \times .976083 = \$1,220.10. \quad \textit{Ans.}$$

EXAMPLE. — For what sum must I make a note having 6 months to run, in order that the avails at bank, if discounted on the day of the date of the note, may amount to $956.38, interest being 6 per cent. per annum?

$$\text{By the table, } \$956.38 \div .9695 = \$986.47. \quad \textit{Ans.}$$

EXAMPLE. — What is the rate of bank interest when the nominal or legal rate is 7 per cent.?

$$.07 \div (1 - .07) = .07527 = 7\tfrac{1}{2} + \tfrac{27}{1000} \text{ per cent.} \quad \textit{Ans.}$$

NOTE. — A note having 5 months to run from Feb. 1 will fall due July 1–4; and the time, including grace, is 5 mo. 3 ds. = 155 days, mean time. But the time in days from Feb. 1 to July 4, when February has but 28 days, is 153 days only, or 2 days short of mean time. See SEC. B., p. 3.

DISCOUNT.

DISCOUNT is a deduction of the interest on the present worth or availability of a debt not yet due, in consideration of its present payment. The *principal* is the present nominal value of the debt, interest included, if any interest has accrued. The *time* is the interval from the present to the date at which the debt will become due. The *rate* is the legal rate of interest, if no other rate is specified; and the *present worth* is that sum of money, which, if put at interest at the same rate and for the same time as the discount, will amount to the principal.

Let a represent the principal, d the discount, w the present worth, and i the interest on one dollar for the time and at the rate of the discount.

$$w = a \div (1 + i) = a - d. \quad d = ai \div (1 + i) = a - w.$$
$$a = d\,(1 + i) \div i = d + w.$$

EXAMPLE. — Required the discount on \$250 for 8 months at 6 per cent.

The interest on \$1 for 8 months at 6 per cent. is .04 of a dollar, or 4 cts.; and

$$250 \times .04 \div (1 + .04) = \$9.6154. \quad \textit{Ans.}$$

EXAMPLE. — Required the present worth of \$1272.62 due 247 days hence, discount 7 per cent.

The interest on \$1 for 247 days at 7 per cent. $= 247 \times .07 \div 365 = 0.04737$, and

$$1272.62 \div 1.04737 = \$1215.06. \quad \textit{Ans.}$$

NOTE. — "*Taking off*, in common parlance, a certain per centum from the face of a demand, is equal to deducting the interest, at that rate per centum, on the present worth for 1 year, plus the interest on the interest of the present worth, at the same rate per centum for 1 year. See SEC. B., p. 1.

COMPOUND DISCOUNT.

COMPOUND DISCOUNT is to compound interest what simple discount is to simple interest. In both cases of discount, the difference between the principal and the discount is that sum of money, which, if put at interest for the same length of time, at the same rate, and in the same general manner as the discount, will amount to the principal.

RULE. — Add 1 to the rate per cent. of the discount for the

working-time, and raise the sum to a power corresponding with the number of working-times; divide the principal by the power, and the quotient will be the present worth; subtract the present worth from the principal, and the remainder will be the compound discount.

NOTE.—The TABLES of the powers of $1 + r$, applicable to compound interest, are equally applicable to compound discount.

EXAMPLE.—Required the present worth of a debt of \$250, allowing yearly compound discount, at 7 per cent. a year, for 3 years 84 days.

$1 + \frac{.07 \times 84}{365} = 1.01611$, the working-rate for the 84 days, and

$$250 \div (1.07^3 \times 1.01611) = \$200.84. \quad \textit{Ans.}$$

EXAMPLE.—What is the present worth of a debt of \$150.25, due 3 years, 3 months, and 10 days hence, without interest, allowing compound discount per quarter-year, at $1\frac{1}{2}$ per cent. per quarter-year?

$$150.25 \div \left(1.015^{13} \times 1.\frac{.06 \times 10}{365}\right) = \textit{Ans.}$$

By table, $150.25 \div (1.19562 \times 1.015 \times 1.00164) =$ \$123.61. *Ans.*

NOTE.—What is here denominated the debt, or principal, represents the debt at the close of the time of the discount; that is, if the debt be on interest, the interest must be included in what is here called the debt, or principal.

PROFIT AND LOSS.

The term "PROFIT AND LOSS," as intimated in treating of PERCENTAGE, relates to the positive and negative interests in business, and embraces the idea of both.

Both profit and loss are absolute quantities, and are expressed by the difference of the cost price and selling price that limit them. They are usually, however, estimated by percentage, predicated upon the first-mentioned price or prime cost.

When the selling price is greater than the cost price, or when the money obtained by the disposal of property exceeds what the property cost, the difference is positive, and denotes increase, profit, or gain. Conversely, when the cost price is greater than the selling price, or when property is disposed of for less money than it cost, the difference is negative, and denotes decrease, loss, or waste. So, the difference of the two prices, divided by the cost price, expresses the rate of gain on the cost when the selling price is the greater, — expresses the rate of loss on the cost when the cost price is the greater.

Let c represent the cost price, purchase price, par value, or sum of money paid for the property; s, the selling price, trade price, premium price, or sum of money received in exchange for the property; r, the rate of the profit or loss; p, the rate per cent. of the profit or loss.

To find the rate or rate per cent. of the profit or loss.

$r = \frac{s \sim c}{c}$. $p = \frac{(s \sim c)\,100}{c}$. Moreover, when the difference is positive, $r = \frac{s}{c} - 1$; and, when it is negative, $r = 1 - \frac{s}{c}$.

Example. — Paid $4 for an article, and sold it for $5. What per cent. was gained? 5 is more than 4 by what per cent. of 4? The difference of 5 and 4 is what per cent. of 4? $5 - 4 = \$1$, gained; and $\frac{5 \sim 4}{4} = .25 = \frac{5}{4} - 1$. 25 per cent. *Ans.*

Example. — Paid $5 for an article, and sold it for $4. What per cent. was lost? 4 is less than 5 by what per cent. of 5? The difference of 4 and 5 is what per cent. of 5? $4 - 5 = -1 = \$1$, lost; and $\frac{5 \sim 4}{5} = .20 = 1 - \frac{4}{5}$. 20 per cent. *Ans.*

Example. — A whistle that cost 3 cents was sold for 20 cents! The profit was how much per cent? $(20 \sim 3) \div 3 = 5\frac{2}{3}$ or $566\frac{2}{3}$ per cent. *Ans.*

Example. — A fop paid $10 for a well-made and well-fitting pair of boots for his own wear, that were worth what they cost him; but, being told that they were unfashionably large, sold them for $4. His vanity cost him what per cent. of the purchase price? $1 - \frac{4}{10} = .6$ or 60 per cent. *Ans.*

To find a price long a given per cent. of the cost, or to find a selling price that shall be the sum of the cost price and a given per cent. of it.

$$s = c + cr = c\,(1 + r) = c\,(100 + p) \div 100.$$

Example. — At what price must I sell an article that cost $2.35 to gain 25 per cent.? 2.35, more 25 per cent. of it, is how much? The sum of $2.35 and 25 per cent. of it is how much? $2.35 + 2.35 \times .25 = 2.35 \times 1.25 = \$2.93\frac{3}{4}$. *Ans.*

To find a price short a given per cent. of the cost, or to find a selling price that shall be the difference of the cost price and a given per cent. of it.

$$s = c - cr = c\,(1 - r) = c\,(100 - p) \div 100.$$

EXAMPLE. — I have a damaged article of merchandise that cost $2.75, and I wish to mark it for sale at 30 per cent. below cost. At what price shall I mark it? 2.75 less 30 per cent. of it is how much? The difference of $2.75 and 30 per cent. of it is how much? 2.75 (1 — .30) = 2.75 × .7 = $1.925. *Ans.*

To find the cost price when the selling price and profit per cent. are given.

$$s = c + cr = c\,(1 + r) \;\therefore\; c = s \div (1 + r) = 100\,s \div (100 + p).$$

EXAMPLE. — What cost that article whose selling price, $4, is long 25 per cent. of the cost? What price, more 25 per cent. of it, is equal to $4? $4 is the sum of what price and 25 per cent. of it? 400 ÷ 125 = $3.20. *Ans.*

To find the cost price when the selling price and loss per cent. are given.

$$s = c - cr = c\,(1 - r) \;\therefore\; c = s \div (1 - r) = 100\,s \div (100 - p)$$

EXAMPLE. — What cost that article whose selling price, $375, is short 7 per cent. of the cost? What price less 7 per cent. of it is equal to $375? $375 is the difference of what price and 7 per cent. of it?

375 ÷ (1 — .07) = 375 ÷ .93 = 375 × 100 ÷ (100 — 7) = $403.226. *Ans.*

EQUATION OF PAYMENTS.

THE EQUATION OF PAYMENTS, or *Averaging of Accounts*, as it is more frequently called, practically consists in finding the common time of maturity of two or more debts due at different times, and is either special or general; special when it is made in regard to a given interchangeable rate of interest and discount, in which the magnitude of the rate slightly affects the time, since discount consumes more time per dollar, rate for rate, than interest; and general, when it is made disrespectful of rate, or common in the greatest possible degree to all rates.

RULE, for common purposes. — Multiply each debt by the number of days between its own date of maturity and that of the debt earliest due, and divide the sum of the products by the sum of the debts; the quotient will express the common time in days subsequent to the leading date.

The following exhibits the face of an account in the ledger, and the time (date) at which it averages due is required.

1860,	April 10 ——	$250.26 —	6 mo.	Due	Oct. 10.
"	June 25 ——	320.56 —	6 "	"	Dec. 25.
"	July 10 ——	50.62 —	3 "	"	Oct. 10.
"	Aug. 1 ——	210.84 —	4 "	"	Dec. 1.
"	" 18 ——	73.40 —	5 "	"	Jan. 18.
"	Oct. 15 ——	100. —	cash	"	Oct. 15.

EXAMPLE. — Practical method of stating and working.

1860.	Due Oct. 10,	$301		
"	" Dec. 25,	321	× 76 =	24396.
"	" " 1,	211	× 52 =	10972.
"	" Jan. 18,	73	× 100 =	7300.
"	" Oct. 15,	100	× 5 =	500.
		1006	)	43168 (43 days, = Nov. 22, 1860. *Ans.*

COMPOUND AVERAGE.

COMPOUND AVERAGE consists in finding the time at which the *balance* of an account or demand averages due, whose sides — the debit and the credit — average due at different dates.

RULE. — Multiply the less sum or side by the difference in days between the two dates — that at which the debit side averages due and that at which the credit side averages due — and divide the product by the difference of the sums or sides; the quotient will be the number of days that one of the dates must be set back, or the other forward, to mark the time sought; for which last,

SPECIAL RULE.

Earlier date with larger sum, set back from earlier.
Later date with larger sum, set forward from later.

EXAMPLE. — The debit side of an account in the ledger foots up $400, and averages due Oct. 12, 1860; the credit side of the same account foots $300, and averages due Nov. 16, 1860. At what date does the balance or difference between the two sides average due?

400   300
300   35
100) 10500 (105 days earlier than Oct. 12, = June 29, 1860. *Ans.*

EXAMPLE. — The debit side of an obligation foots $250, and averages due May 17, 1860; the credit side of the same obligation foots $175, and averages due May 1, 1860. At what date does the difference of the sides average due?

250   175
175   16
75) 2800 (37⅓ days later than May 17, = June 23, 1860. *Ans.*

GENERAL AVERAGE.

It is the established usage that whatever of either of the three commercial interests — the ship, the cargo, or the freight — is voluntarily sacrificed or destroyed for the general good, or with the view of saving the most that may be saved when all is in imminent danger of being lost, is matter of general loss to the respective interests, and not more especially to the interest voluntarily abandoned than to the others. So, too, the losses and damages incident to the voluntary sacrifice, and collateral therewith, together with the expenditures which the master has been compelled to make for the general good, in consequence of disaster, are matters of general average, or are to be contributed for, *pro rata*, by the several interests.

The contributory interests are the ship, the cargo, and the freight, at their net values, independent of charges, premiums paid for insurance, &c.

The contributory value of the ship, generally, is her value at the port of departure at the time of leaving, less the premium paid for her insurance.

The contributory value of the cargo is its net value, in a sound state, at the port of destination, if the voyage be completed; or its invoice value if the voyage be broken up and the cargo returned to the port whence it was shipped; or its market-value at any intermediate port, where of necessity it is discharged and disposed of. The value of the goods jettisoned, and to be contributed for, is their value after the same manner; and that value is a part of the contributory value of the cargo, as well as a matter of general average.

The contributory value of the freight, generally, is the gross amount or amount per freight-list, less one-third part thereof, in most of the States; but, in the State of New York, one-half thereof, for seamen's wages and other expenses. The loss of freight by jettison, when any freight is earned, is matter of general average. If the cargo is transshipped on board another vessel, and in that way sent to the port of destination, the contributory value of the freight is the gross amount, less the sum paid the other vessel.

The voluntary damage to the ship, with a view to the general good, — such as throwing over her furniture, destroying her equipments, cutting away her masts, breaking up her decks to get at the cargo for the purpose of throwing it over, &c., — is contributed for at two-thirds the cost of repairing and restoring; the new articles being supposed one-half better, or worth one-half more, than the old.

If we let $V =$ contributory value of the vessel,
$C =$ contributory value of the cargo,
$F =$ contributory value of the freight,
$d =$ aggregate amount of losses to be averaged, then $d \div (V + C + F) = r$, the per cent. of each interest that each must contribute, and

$V \times r =$ Vessel's share of the loss,
$C \times r =$ Cargo's share of the loss,
$F \times r =$ Freight's share of the loss.

When a contributory interest's share of the loss is to be distributed among the several owners of that interest, the same *pro rata* method is to be observed: thus

$A \times r =$ sum A must contribute,
$B \times r =$ sum B must contribute,
$D \times r =$ sum D must contribute;

A, B, and D being A's, B's, and D's respective shares in that interest.

ASSESSMENT OF TAXES.

$G =$ amount of taxable property, real and personal, as per grand list.
$A =$ amount of money to be raised, including the whole poll-tax.
$T =$ amount of money to be raised on property alone.
$n =$ number of ratable polls.
$h =$ poll-tax per head.
$r =$ rate per cent. to be raised on taxable property.
$P =$ an individual's taxable property, as per grand list.
$b =$ P's poll-tax.

$T = A - hn$. $r = T \div G$. $Pr + b =$ P's tax, including poll.

INSURANCE.

Insurance is a written contract of indemnity, called the *policy*, by which one party (the *insurer* or *underwriter*) engages, for a stipulated sum, called the *premium* (usually a per cent. on the value of the property insured), to insure another against a risk or loss to which he is exposed.

Let $P =$ Principal, or amount insured on,
$r =$ rate per cent. of insurance,
$a =$ premium for insurance.

$$a = Pr. \quad r = a \div P. \quad P = a \div r.$$

Example.—What is the premium for insuring on \$4500 at $1\frac{1}{2}$ per cent.?

$$4500 \times .015 = \$67.50. \quad \textit{Ans.}$$

LIFE-INSURANCE.

Life-insurance is predicated upon the even chance in years, called the *expectation of life*, that an individual in general health at any given age appears by the rates of mortality to have of living beyond that age.

The Carlisle Tables of Expectation, column C in the following tables, are used almost or quite exclusively in England, and by some insurance-companies in the United States; while those by Dr. Wigglesworth, column W, computed with special reference to the rates of mortality in this country, are used by others.

The Supreme Court of Massachusetts has adopted the Wigglesworth rates of expectation in estimating the value of life-annuities and life-estates.

TABLE

Of Ages and Expectations from Birth to 103 Years.

Age.	C.	W.	Age.	C.	W.	Age.	C.	W.	Age.	C.	W.
0	38.72	28.15	26	37.14	31.93	52	19.68	20.05	78	6.12	6.59
1	44.68	36.78	27	36.41	31.50	53	18.97	19.46	79	5.80	6.21
2	47.55	38.74	28	35.69	31.08	54	18.28	18.92	80	5.51	5.85
3	49.82	40.01	29	35.00	30.66	55	17.58	18.35	81	5.21	5.50
4	50.76	40.73	30	34.34	30.25	56	16.89	17.78	82	4.93	5.16
5	51.25	40.88	31	33.68	29.83	57	16.21	17.20	83	4.65	4.87
6	51.17	40.69	32	33.03	29.43	58	15.55	16.63	84	4.39	4.66
7	50.80	40.47	33	32.36	29.02	59	14.92	16.04	85	4.12	4.57
8	50.24	40.14	34	31.68	28.62	60	14.34	15.45	86	3.90	4.21
9	49.57	39.72	35	31.00	28.22	61	13.82	14.86	87	3.71	3.90
10	48.82	39.23	36	30.32	27.78	62	13.31	14.26	88	3.59	3.67
11	48.04	38.64	37	29.64	27.34	63	12.81	13.66	89	3.47	3.56
12	47.27	38.02	38	28.96	26.91	64	12.30	13.05	90	3.28	3.43
13	46.51	37.41	39	28.28	26.47	65	11.79	12.43	91	3.26	3.32
14	45.75	36.79	40	27.61	26.04	66	11.27	11.96	92	3.37	3.12
15	45.00	36.17	41	26.97	25.61	67	10.75	11.48	93	3.48	2.40
16	44.27	35.76	42	26.34	25.19	68	10.23	11.01	94	3.53	1.98
17	43.57	35.37	43	25.71	24.77	69	9.70	10.50	95	3.53	1.62
18	42.87	34.98	44	25.09	24.35	70	9.18	10.06	96	3.46	
19	42.17	34.59	45	24.46	23.92	71	8.65	9.60	97	3.28	
20	41.46	34.22	46	23.82	23.37	72	8.16	9.14	98	3.07	
21	40.75	33.84	47	23.17	22.83	73	7.72	8.69	99	2.77	
22	40.04	33.46	48	22.50	22.27	74	7.33	8.25	100	2.28	
23	39.31	33.08	49	21.81	21.72	75	7.01	7.83	101	1.79	
24	38.59	32.70	50	21.11	21.17	76	6.69	7.40	102	1.30	
25	37.86	32.33	51	20.39	20.61	77	6.40	6.99	103	0.83	

Thus, by the tables, a man in general good health at 21 years of age has an even chance, by the Carlisle rate of mortality, of living $40\frac{3}{4}$ years longer; by the Wigglesworth rate, of living $33\frac{84}{100}$ years longer. So a man in general good health, at 60 years of age, has, by the Carlisle rate, an even chance of living 14.34 years longer; by the Wigglesworth rate, an even chance of living 15.45 years longer, etc.

FELLOWSHIP.

FELLOWSHIP calls for the distribution of a given effect to each of the several causes associated in its production, proportional to their respective magnitudes one with another.

It is a rule, therefore, adapted to the use of partners associated in business, in achieving a *pro rata* distribution among themselves as individuals, of the profits or losses pertaining to the company.

RULE.—Multiply each partner's investment or share of the capital stock, by the whole gain or loss, and divide the product by the sum of all the shares, or gross capital.

EXAMPLE.—Three men, A, B, and C, enter into partnership. A invests \$500, B \$700, and C \$300. They trade and gain \$400. What is each partner's share of the profits?

A, \$500	$500 \times 400 \div 1500 =$ \$133.33⅓ = A's share.
B, 700	$700 \times 400 \div 1500 =$ 186.66⅔ = B's "
C, 300	$300 \times 400 \div 1500 =$ 80.00 = C's "
\$1500 = gross capital.	\$400.00 Proof.

EXAMPLE.—D's investment of \$600 has been employed eight months; E's, of \$500, five months; and F's, of \$300, five months, the profits of the company are \$500, and are to be divided *pro rata* among the partners. What is each partner's share?

D, $\$600 \times 8 = 4800 \times 500 \div 8800 =$ \$272.73, D's share.
E, $500 \times 5 = 2500 \times 500 \div 8800 =$ 142.05, E's "
F, $300 \times 5 = 1500 \times 500 \div 8800 =$ 85.22, F's "
8800 — \$500. Proof.

EXAMPLE.—Of \$120 distributed, there were given to A, $\frac{1}{3}$; to B, $\frac{1}{4}$; to C, $\frac{1}{5}$; and to D, $\frac{1}{6}$, and there was nothing remaining. What sum did each receive?

$\frac{1}{3}$ of $120 = 40 \times 120 \div 114 =$ \$$42\frac{2}{19}$ = A's share.
$\frac{1}{4}$ of $120 = 30 \times 120 \div 114 =$ $31\frac{11}{19}$ = B's "
$\frac{1}{5}$ of $120 = 24 \times 120 \div 114 =$ $25\frac{5}{19}$ = C's "
$\frac{1}{6}$ of $120 = 20 \times 120 \div 114 =$ $21\frac{1}{19}$ = D's "
114 — \$120. Proof.

EXAMPLE.—Divide the number 180 into 3 parts, which shall be to each other as $\frac{1}{2}$, $\frac{1}{3}$, $\frac{1}{4}$.

$\frac{1}{2}$ of $180 = 90 \times 180 \div 195 = 83.08$
$\frac{1}{3}$ of $180 = 60 \times 180 \div 195 = 55.38$
$\frac{1}{4}$ of $180 = 45 \times 180 \div 195 = 41.54$
195 — 180.00 Proof.

EXAMPLE.—$400 are to be divided between A, B, and C, in the ratio of ½ to A, ½ to B, and ¼ to C; how much will each receive?

$\frac{1}{2}$ of 400 = 200, and 200 × 400 ÷ 500 = $160 = A's share.
$\frac{1}{2}$ of 400 = 200, and 200 × 400 ÷ 500 = 160 = B's share.
$\frac{1}{4}$ of 400 = 100, and 100 × 400 ÷ 500 = 80 = C's share.
500 $400. Proof.

ALLIGATION.

ALLIGATION *Medial* is a method by which to find the mean price of a mixture or compound, consisting of two or more articles or ingredients, the quantity and price of each being given.

RULE.—Multiply each quantity by its price, and divide the sum of the products by the sum of the quantities; the quotient will be the price per unity of measure of the mixture; and, having found the price of the given quantities as mixed, any quantities of the same materials, taken in like proportions, will be at the same price.

EXAMPLE.—If 20 lbs. of sugar at 8 cents, 40 lbs. at 7 cents, and 80 lbs. at 5 cents per pound, be mixed together, what will be the mean price, or price per pound, of the mixture?

20 × 8 = 160
40 × 7 = 280
80 × 5 = 400
140) 840 (6 cents. *Ans.*

The several kinds, then, at their respective prices, taken in the proportion of 1 at 8, 2 at 7, and 4 at 5 cts., will form a mixture worth 6 cts. a pound.

EXAMPLE.—If 10 lbs. of nickel are worth $2, and 24 lbs. of copper are worth $4½, and 8 lbs. of zinc are worth 40 cts., and 1 lb. of lead is worth 5 cts., what are 5 lbs. of *pretty good* German silver worth?

$$\frac{(200+450+40+5)\times 5}{43} = 81 \text{ cents. } \textit{Ans.}$$

ALLIGATION *Alternate* is a method by which to find what quantity of each of two or more articles or ingredients, whose prices or qualities are given, must be taken to form a mixture or compound that shall be at a given price or of a given quality between the two extremes. It also applies to the finding of relative quantities when the quantity of one or more of the articles is limited.

RULE.—Connect the given prices or qualities—a less than the given mean with that one or either one that is greater—and to the extent that all be thus connected; then place the difference between each given and the given mean opposite, not the given, or the given mean, but the given with which it is alligated; the number standing opposite each price or quality will be the quantity that must be taken at that price, or of that quality, to form a mixture or compound at the price or of the quality desired. And, being proportions respectively to each other, they may be taken in ratio greater or less, as desired. See Sec. B, page 18 *b*.

Example. — In what proportions shall I mix teas at 48 cents a pound and 54 cents a pound, that the mean price may be 50 cents a pound?

In the proportions

$$50\left\{\begin{array}{l}48\\54\end{array}\right\}\left\{\begin{array}{l}\text{4 lbs at 48 cts.}\\\text{2 lbs. at 54 cts.}\end{array}\right\}\textit{Ans.}$$

Or, as 2 at 48 to 1 at 54.

$$\text{Proof.}\left\{\begin{array}{l}\overline{2\times 48}+\overline{1\times 54}=150.\\3\times 50\qquad\qquad=150.\end{array}\right.$$

Example. — In what proportions shall I mix teas at 48, 54, and 72 cents a pound, that the mixture may average 60 cents a pound?

$$60\left\{\begin{array}{l}48\\54\\72\end{array}\right.\quad\begin{array}{l}12,\\12,\\12+6,\end{array}\quad\left.\begin{array}{l}\text{12 at 48}\\\text{12 at 54}\\\text{18 at 72}\end{array}\right\}=\left\{\begin{array}{l}\text{2 at 48}\\\text{2 at 54}\\\text{3 at 72}\end{array}\right\}\textit{Ans.}$$

Example. — A wine dealer has received an order for a quantity of wine at 50 cts. a gallon. He has none ready *manufactured* at that price. He has it at 40 cts., at 56 cts., and at 80 cents a gallon, and he has water that cost him nothing. He wishes to fill the order with a mixture composed of the four materials — the water and the three different priced wines. In what proportions may he mix them, that the mean or average price shall be 50 cents a gallon?

$$50\left\{\begin{array}{l}00\\40\\56\\80\end{array}\right.\overset{\textit{Ans.}}{\left.\begin{array}{r}6\\30\\50\\\underline{10}\end{array}\right\}}\quad\text{Or, }50\left\{\begin{array}{l}00\\40\\56\\80\end{array}\right.\begin{array}{l}30\\6+30\\10\\50+10\end{array}\overset{\textit{Ans.}}{\left.\begin{array}{r}=30\\=36\\=10\\=\underline{60}\end{array}\right\}}$$

$$=96\text{ gals.}\qquad\qquad=136\text{ gals.}$$

$$\text{Or, }50\left\{\begin{array}{l}00\\40\\56\\80\end{array}\right.\overset{\textit{Ans.}}{\left.\begin{array}{l}6+30\\30\\50\\\underline{50+10}\end{array}\right\}}\quad\text{Or, }50\left\{\begin{array}{l}00\\40\\56\\80\end{array}\right.\overset{\textit{Ans.}}{\left.\begin{array}{l}6\\6+30\\50+10\\\underline{10}\end{array}\right\}}$$

$$=176\text{ gals.}\qquad\qquad=112\text{ gals.}$$

If, now, having found the proportions desired, it is wished to limit one of the articles in quantity — say the best wine to 8 gallons in the mixture — the proportions of the remaining articles thereto are found thus: —

Instance, 1st example, —

$$\left.\begin{array}{l}10 : 8 :: 50 = 40 \\ 10 : 8 :: 30 = 24 \\ 10 : 8 :: \ 6 = 4\frac{4}{5}\end{array}\right\} \begin{array}{l}\text{And the mixture will consist of} \\ 8 + 40 + 24 + 4\frac{4}{5} = 76\frac{4}{5} \text{ gallons.}\end{array}$$

If, instead, it is desired to mix a given quantity, say 100 gallons, and proportioned, say as in first example, the quantity to be taken of each is ascertained by the following

RULE. — As the sum of the relative quantities is to the quantity required, so is each relative quantity to the quantity required of it respectively.

The sum of the relative quantities alluded to is $6 + 30 + 50 + 10 = 96$; then,

$$\begin{array}{l}96 : 100 :: 6 = 6\frac{1}{4} \\ 96 : 100 :: 30 = 31\frac{1}{4} \\ 96 : 100 :: 50 = 52\frac{1}{12} \\ 96 : 100 :: 10 = 10\frac{5}{12}\end{array}$$

INVOLUTION.

INVOLUTION consists in involving, that is, in multiplying a number one or more times into itself. The number so involved is called the *root*, and the product arising from such involution, its *power*.

The *second power*, or *square*, of the root, is obtained by multiplying the root *once* into itself, as $4 \times 4 = 16$; 4 being the root and 16 its square.

The *third power*, or *cube*, of a number, is obtained by multiplying the number twice into itself, as $4 \times 4 \times 4 = 64$; and so on for any power whatever.

When a number is to be involved into itself, a small figure called the *index* or *exponent* is placed at its right, indicating the number of times it is to be so involved, or the power to which it is to be raised. Thus, $3^4 = 3 \times 3 \times 3 \times 3 = 81$; and $4^3 = 4 \times 4 \times 4 = 64$.

EVOLUTION.

EVOLUTION is the opposite of Involution. It consists in finding a root of a given number, instead of a power of a given root.

When the root of a number is required or indicated, the number is written with the $\sqrt{}$ before it: and the character or denomination of the root, if it be other than the square root, is defined by an index figure placed over the sign. When the square root of a number is required, the sign ($\sqrt{}$) is placed before the number, but the index (2) is usually omitted. Thus, $\sqrt{25}$, shows that the square root of 25 is required, or to be taken ; and $\sqrt[3]{25}$ shows that the cube root is required. The operation is usually called extracting the root.

TO EXTRACT THE SQUARE ROOT.

RULE — 1. Separate the given number into periods of two figures each, by placing a point over the *first* figure, *third*, *fifth*, &c., counting from right to left — the root will consist of as many figures as there are periods.

2. Find the greatest square in the left hand period, and place its root in the quotient; subtract the square of the root from the left hand period, and to the remainder bring down the next period for a dividend.

3. Multiply the root so far found — the figure in the quotient — by 2, for a divisor; see how many times the divisor is contained in the dividend, except the right hand figure, and place the result (the number of times it is contained) in the quotient, to the right of the figure already there, and also to the right of the divisor ; multiply the divisor, thus increased, by the last figure in the quotient, and subtract the product from the dividend, and to the remainder bring down the next period for a dividend.

4. Multiply the quotient — the root so far found (now consisting of two figures) — by 2, as before, and take the product for a divisor ; see how many times the divisor is contained in the dividend, except the right hand figure, and place the result in the quotient, and to the right of the divisor, as before ; multiply the divisor, as it now stands, by the figure last placed in the quotient, and subtract the product from the dividend, and to the remainder bring down the next period for a dividend, as before.

5. Multiply the quotient (now consisting of 3 figures) by 2, as before, and take the product for a divisor, and in all respects proceed as when seeking for the last two figures in the quotient. The quotient, when all the periods have been brought down and divided, will be the root sought.

NOTE. — 1. If there is a remainder after finding the integer of a root, annex periods of ciphers thereto, and proceed as when seeking for the integer. The quotient figures will be the decimal portion of the root.

2. If the given number is a decimal, or consists of a whole number and decimal, point off the decimal from left to right, by placing the point over the *second*, *fourth*, *sixth*, &c., figures therein, and fill the last period, if incomplete, by annexing a cipher.

3. If the dividend does not contain the divisor, a cipher must be placed in the quotient, and also at the right of the divisor, and the next period brought down ; then the dividend must be divided by the divisor as increased.

4. If the quotient figure, obtained by dividing by the double of the root, is too large, as will sometimes be the case, (see 3d Example) it must be dropped, and a less — one which is the true measure — taken in its stead.

EXAMPLE. — Required the square root of 123456.432.

```
        123456.4320 ( 351.3636+.  Ans.
        9
   65 ) 334
        325
    70 . ) 956
           701
    7023 ) 25543
           21069
    70266 ) 447420
            421596
    702723 ) 2582400
             2108169
    7027266 ) 47423100
              42163596
               5259504
```

EXAMPLE. — Required the square root of 10621. Also, of 28561

```
     10621 ( 103.05+.  Ans.
     1
203 ) 00621
        609
20605 ) 120000
        103025
         16975
```

```
     28561 ( 169.  Ans.
     1
26 ) 185
     156
329 ) 2961
      2961
```

TO EXTRACT THE CUBE ROOT.

RULE — 1. Separate the given number into periods of three figures each, by placing a point over the *first*, *fourth*, *seventh*, &c., counting from right to left — the root will consist of as many figures as there are periods.

2. Find the greatest cube in the left hand period, and place its root in the quotient; subtract the cube of the root from the left hand period, and to the remainder bring down the next period for a dividend.

3. Multiply the square of the quotient by 300, for a divisor; see how many times the divisor is contained in the dividend, and place the result (except that the remainder is large, diminished by one or two units) in the quotient.

4. Multiply the divisor by the figure last placed in the quotient, and to the product add the square of the same figure, multiplied by the other figure, or figures, in the quotient, and by 30; and add also thereto

the cube of the same figure, and take the sum for the subtrahend; subtract the subtrahend from the dividend, and to the remainder bring down the next period for a dividend, with which proceed as with the preceding, so continuing until the whole is completed.

NOTE — 1. Decimals must be pointed from left to right, by placing a point over the *third*, *sixth*, &c., figures in that direction.

2. If the divisor is not contained by the dividend, place a cipher in the quotient, and annex two ciphers to the divisor, and bring down the next period for a dividend, and use the divisor, as thus increased, for finding the next quotient figure.

3. If there is a remainder after finding the integer of the root, annex a period of three ciphers thereto, and proceed for the decimal of the root as if seeking for the integer, annexing a period of three ciphers to each remainder until the decimal is carried to as many places of figures as desired.

EXAMPLE. — Required the cube root of 47421875.6324.

47421875.632400 (361.959+. *Ans.*
27

$3^2 \times 300 = 2700$) 20421
6

16200
$6^2 \times 3 \times 30 = 3240$
$6^3 = 216 = 19656$

$36^2 \times 300 = 388800$) 765875
1

388800
$1^2 \times 36 \times 30 = 1080$
$1^3 = 1 = 389881$

$361^2 \times 300 = 39096300$) 375994632
9

351866700
$9^2 \times 361 \times 30 = 877230$
$9^3 = 729 = 352744659$

$3619^2 \times 300 = 3929148300$) 23249973400
5

19645741500
$5^2 \times 3619 \times 30 = 2714250$
$5^3 = 125 = 19648455875$

$36195^2 \times 300 = 393023407500$) 3601517525000
9

3537210667500
$9^2 \times 36195 \times 30 = 87953850$
$9^3 = 729 = 3537298622079$

64218902921

EXAMPLE.—Required the cube root of 32768. Also, of 8489664.

32768(32. *Ans.*
27

$3^2 \times 300 = 2700$) 5768
2
5400
$2^2 \times 3 \times 30 = 360$
$2^3 = 8 = 5768$

8489664(204. *Ans.*
8

$2^2 \times 300 = 120000$) 489664
4
480000
$4^2 \times 20 \times 30 = 9600$
$4^3 = 64 = 489664$

General Rule for extracting the roots of all powers, or for finding any proposed root of a given number.

1. Point off the given number into periods of as many figures each, counting from right to left, as correspond with the denomination of the root required; that is, if the cube root be required, into periods of *three* figures, if the fourth root, into periods of *four* figures, &c.

2. Find the first figure of the root by inspection or trial, and place it at the right of the number, in the form of a quotient; raise this quotient figure to a power corresponding with the denomination of the root sought, and subtract that power from the left hand period, and to the remainder bring down the first figure of the next period, for a dividend.

3. Raise the root thus far found (the quotient figure) to a power next inferior in denomination to that of the root required, multiply this power by the number or index figure of the root required, and take the product for a divisor; find the number of times the divisor is contained in the dividend, and place the result (except that the remainder is large, diminished by one or two units) in the quotient, for the second figure of the root.

4. Raise the root thus far found (now consisting of two figures) to a power corresponding in denomination with the root required, and subtract that power from the two left hand periods, and to the remainder bring down the first figure of the third period, for a dividend; find a new divisor, as before, and so proceed until the whole root is extracted.

EXAMPLE.—Required the fifth root of 45435424.

45435424(34. *Ans.*
$3^5 = 243$.
$3^4 \times 5$.) 2113
$34^5 = 45435424$

EXAMPLE. — Required the fifth root of 432040.0354.

$$\dot{4}3204\dot{0}.0354\dot{0} \; (\, 13.4 + \quad \textit{Ans.}$$

$$1^5 = 1$$

$$1^4 \times 5) \; 33$$

$$13^5 = 371293$$

$$13^4 \times 5) \; 607470$$

$$13.4^5 = 43204003424$$

$$\cdots\cdots\cdots 116$$

For instructions touching special cases, see NOTES relative to the extraction of the square root, and to the extraction of the cube root.

The $\sqrt{}$ of the $\sqrt{}$ of any number $= \sqrt[4]{}$ of that number
" $\sqrt{}$ of the $\sqrt[3]{} = \sqrt[6]{}$.
" $\sqrt{}$ of the $\sqrt{}$ of the $\sqrt{} = \sqrt[8]{}$.
" $\sqrt[3]{}$ of the $\sqrt[3]{} = \sqrt[9]{}$.
" $\sqrt{}$ of the $\sqrt[5]{} = \sqrt[10]{}$, &c.

ARITHMETICAL PROGRESSION.

A series of three or more numbers, increasing or decreasing by equal differences, is called an *arithmetical progression*. If the numbers progressively increase, the series is called an *ascending arithmetical progression;* and if they progressively decrease, the series is called a *descending arithmetical progression.*

The numbers forming the series are called the *terms* of the progression, of which the first and the last are called the *extremes*, and the others the *means.*

The difference between the consecutive terms, or that quantity by which the numbers respectively increase upon each other, or decrease from each other, is called the *common difference.*

Thus, 3, 5, 7, 9, 11, &c., is an ascending arithmetical progression, and 11, 9, 7, 5, 3, is a descending arithmetical progression. In these progressions, in both instances, 11 and 3 are the *extremes*, of which 11 is the *greater extreme*, and 3 is the *less extreme.* The numbers between these, (9, 7, 5,) are the *means.*

In every arithmetical progression, the sum of the extremes is equal to the sum of any two means that are equally distant from the extremes; and is, therefore, equal to *twice* the middle term, when the series consists of an odd number of terms. Thus, in the foregoing series, $3 + 11 = 5 + 9 = 7 \times 2$.

The *greater extreme*, the *less extreme*, the *number of terms*, the *common difference*, and the *sum of the terms*, are called the *five* properties of an arithmetical progression, of which, any *three* being given, the other two may be found.

Let s represent the sum of the terms.
" E " the greater extreme.
" e " the less extreme.
" d " the common difference.
" n " the number of terms.

The extremes of an arithmetical progression and the number of terms being given, to find the sum of the terms.

$$\frac{(E+e)\times n}{2} = \text{sum of the terms.}$$

EXAMPLE. — What is the sum of all the even numbers from 2 to 100, inclusive?

$$102 \times 50 \div 2 = 2550. \quad \textit{Ans.}$$

EXAMPLE. — How many times does the hammer of a common clock strike in 12 hours?

$$(1 + 12) \times 12 \div 2 = 78 \text{ times.} \quad \textit{Ans.}$$

$$\left(\frac{E-e}{d}+1\right)\times\frac{E+e}{2} = \text{sum of the terms.}$$

$$(E \times 2 - \overline{\overline{n-1} \times d}) \times \tfrac{1}{2}\, n = \text{sum of the terms.}$$

$$(2\,e + \overline{\overline{n-1} \times d}) \times \tfrac{1}{2}\, n = \text{sum of the terms.}$$

The greater extreme, the common difference, and the number of terms of an arithmetical progression being given, to find the less extreme.

$$E - (d \times \overline{n-1}) = \text{less extreme.}$$

EXAMPLE. — A man travelled 18 days, and every day 3 miles farther than on the preceding; on the last day he travelled 56 miles; how many miles did he travel the first day?

$$56 - (\overline{18-1} \times 3) = 5 \text{ miles.} \quad \textit{Ans.}$$

$$\frac{s}{n} - \left(\frac{\overline{n-1} \times d}{2}\right) = \text{less extreme.}$$

$$\frac{s}{n} \times 2 - E = \text{less extreme.}$$

$$\frac{\sqrt{(E\times 2+d)^2-s\times d\times 8}+d}{2}=\text{less extreme, when}$$

$\sqrt{(2E+d)^2-8sd}$ is equal to, or greater than d.

$$\frac{\sqrt{(2E+d)^2-8sd}\backsim d}{2}=\text{less extreme, when}$$

$\sqrt{(2E+d)^2-8sd}$ is less than d.

$$\frac{\sqrt{(2e\backsim d)^2+8sd}-d}{2}=\text{greater extreme}$$

$$d\times\overline{n-1}+e=\text{greater extreme.}$$

$$\frac{s}{n}+\frac{\overline{n-1}\times d}{2}=\text{greater extreme.}$$

$$2s\div n-e=\text{greater extreme.}$$

The extremes of an arithmetical progression and the common difference being given, to find the number of terms.

$$E-e\div d+1=\text{number of terms.}$$

EXAMPLE. — As a heavy body, falling freely through space, descends $16\frac{1}{12}$ feet in the first second of its descent, $48\frac{3}{12}$ feet in the next second, $80\frac{5}{12}$ in the third second, and so on; how many seconds had that body been falling, that descended $305\frac{7}{12}$ feet in the last second of its descent?

$305\frac{7}{12}-16\frac{1}{12}=289\frac{1}{2}\div 32\frac{1}{6}=9+1=10$ seconds. *Ans.*

$$\frac{\sqrt{(2e\backsim d)^2+8sd}-d}{2}-e\div d+1=\text{number of terms.}$$

$$\overline{2s\div \frac{E+\sqrt{(2E+d)^2-8sd}+d}{2}}=\text{number of terms when}$$

$\sqrt{(2E+d)^2-8sd}$ is equal to, or greater than d.

$$\overline{2s\div \frac{E+\sqrt{(2E+d)^2-8sd}\backsim d}{2}}=\text{number of terms when}$$

$\sqrt{(2E+d)^2-8sd}$ is less than d.

$$\frac{s\times 2}{E+e}=\text{number of terms.}$$

The extremes of an arithmetical progression, and the number of terms being given, to find the common difference.

$$\frac{E - e}{n - 1} = \text{common difference.}$$

EXAMPLE. — One of the extremes of an arithmetical progression is 28 and the other is 100, and there are 19 terms in the series; required the common difference.

$$100 \backsim 28 \div 1 \backsim 19 = 4. \quad \textit{Ans.}$$

$$E - e \div \left(\frac{s \times 2}{E + e} - 1\right) = \text{common difference.}$$

$$\frac{2\,s \div n - 2\,e}{n - 1} = \text{common difference.}$$

$$\frac{2\,E - (2\,s \div n)}{n - 1} = \text{common difference.}$$

EXAMPLE. — The less extreme of an arithmetical progression is 28, the sum of the terms 1216, and the number of terms 19; required the 7th term in the series, descending.

$1216 \times 2 \div 19 = 128 =$ sum of the extremes.
$128 - 28 = 100 =$ greater extreme.
$100 - 28 = 72 =$ difference of extremes.
$72 \div \overline{n - 1}\ (18) = 4 =$ common difference.
$100 - (\overline{7 - 1} \times 4) = 76 =$ 7th term descending. *Ans.*

Required the 5th term from the less extreme, in an arithmetical progression, whose greatest extreme is 100, common difference 4, and number of terms 19.

$$100 - (\overline{19 - 5} \times 4) = 44. \quad \textit{Ans.}$$

To find any assigned number of arithmetical means, between two given numbers or extremes.

RULE. — Subtract the less extreme from the greater, divide the remainder by 1 more than the number of means required, and the quotient will be the common difference between the extremes; which, added to the less extreme, gives the least mean, and, added to that, gives the next greater, and so on.

Or, $\overline{E - e} \div \overline{m + 1} = d$, E being the greater extreme, e the less extreme, m the number of means required, and d the common difference.

And $e + d$, $e + 2\,d$, $e + 3\,d$, &c.; or, $E - d$, $E - 2\,d$, $E - 3\,d$, &c., will give the means required.

13 *

EXAMPLE.—Required to find 5 arithmetical means between the numbers 18 and 3.

$$18 - 3 = 15 \div 6 = 2\tfrac{1}{2}, \text{ and}$$

$$3 + 2\tfrac{1}{2} = 5\tfrac{1}{2} + 2\tfrac{1}{2} = 8 + 2\tfrac{1}{2} = 10\tfrac{1}{2} + 2\tfrac{1}{2} = 13 + 2\tfrac{1}{2} = 15\tfrac{1}{2}.$$

$5\tfrac{1}{2}$, 8, $10\tfrac{1}{2}$, 13, $15\tfrac{1}{2}$, therefore, are 5 arithmetical means, between the extremes, 3 and 18.

NOTE.—The arithmetical mean between any two numbers may be found by dividing the sum of those numbers by 2; thus, the arithmetical mean of 9 and 8 is $(9+8) \div 2 = 8\tfrac{1}{2}$.

GEOMETRICAL PROGRESSION.

A series of three or more numbers, increasing by a common multiplier, or decreasing by a common divisor, is called a *geometrical progression*. If the greater numbers of the progression are to the right, the progression is called an *ascending geometrical progression*, but, on the contrary, if they are to the left, it is called a *descending geometrical progression*. The number by which the progression is formed, that is, the common multiplier, or divisor, is called the *ratio*.

The numbers forming the series are called the *terms* of the progression, of which the first and the last are called the *extremes*, and the others the *means*. The greater of the extremes is called the *greater extreme*, and the less the *less extreme*.

Thus, 3, 6, 12, 24, 48, is an ascending geometrical progression, because 48 is as many times greater than 24, as 24 is greater than 12, &c.; and 250, 50, 10, 2, is a descending geometrical progression, because 2 is as many times less than 10, as 10 is less than 50, &c.

In the first mentioned series, (3, 6, 12, 24, 48,) 48 is the *greater extreme*, and 3 is the *less extreme;* the numbers 6, 12, 24 are the means in that progression.

So, too, of the progression 250, 50, 10, 2; 250 and 2 are the extremes, and 50 and 10 are the means.

In the first mentioned progression, 2 is the ratio, and in the last, or in the progression 2, 10, 50, 250, 5 is the ratio.

In a geometrical progression, the product of the two extremes is equal to the product of any two means that are equally distant from the extremes, and, also, equal to the square of the middle term, when the progression consists of an odd number of terms.

Thus, in the progression 2, 6, 18, 54, 162; $162 \times 2 = 54 \times 6 = 18 \times 18$.

When a geometrical progression has but 3 terms, either of the extremes is called a *third proportional* to the other two; and the middle term, consequently, is a *mean proportional* between them.

Thus, in the progression 48, 12, 3, 3 is a *third proportional* to 48 and 12, because 48 divided by the ratio = 12, and 12 divided by the ratio = 3; or 3 × ratio = 12, and 12 × ratio = 48 : 12 is the *mean proportional*, because 12 × 12 = 48 × 3.

Of the 5 properties of a geometrical progression, viz., the *greater extreme*, the *less extreme*, the *number of terms*, the *ratio*, and the *sum of the terms*, any three being given, the other two may be found

Let s represent the sum of the terms.
" E " the greater extreme.
" e " the less extreme.
" r " the ratio.
" n " the number of terms.
" ^{n} when affixed as an index or exponent, represent that the term, number, or quantity, to which it is affixed, is to be raised to a power equal to the number of terms in the respective progression, &c.

Any three of the five parts of a geometrical progression being given, to find the remaining two parts.

$$\frac{E-e}{r-1}+E = \text{sum of the terms.}$$

$$\frac{E\times r-e}{r-1} = \text{sum of the terms.}$$

$$\frac{r^{n}\times e-e}{r-1} = \text{sum of the terms.}$$

$$\frac{E-(E\div \overline{r^{n-1}})}{r-1}+E = \text{sum of the terms.}$$

$$\frac{E-e}{\sqrt[n-1]{(E\div e)}-1}+E = \text{sum of the terms.}$$

EXAMPLE.—The greater extreme of a geometrical progression is 162, the less extreme is 2, and there are 5 terms in the progression: required the sum of the series.

$$\frac{162-2}{\sqrt[4]{(162\div 2)}-1} = 80+162 = 242. \quad \textit{Ans.}$$

$$\frac{s\times \overline{r-1}+e}{r} = \text{greater extreme.}$$

$$\frac{r^n}{r} \times e = \text{greater extreme.}$$

$$r^{\overline{n-1}} \times e = \text{greater extreme.}$$

$$\frac{s \times r^{\overline{n-1}} \times \overline{r-1}}{r^n - 1} = \text{greater extreme.}$$

$$s - \overline{(s - E) \times r} = \text{less extreme.}$$

$$E \div r^{\overline{n-1}} = \text{less extreme.}$$

$$\frac{s \times r^{\overline{n-1}} \times \overline{r-1}}{r^n - 1} \div r^{n-1} = \text{less extreme.}$$

$$\frac{s - e}{s - E} = \text{ratio.}$$

$$\sqrt[n-1]{\frac{E}{e}} = \text{ratio.}$$

$\frac{s \times \overline{r-1}}{e} + 1 = r^n$; n, therefore, is equal to the number of times that r must be multiplied into itself to equal $\frac{s \times \overline{r-1}}{e} + 1$.

$$\frac{s \times \overline{r-1}}{s - \overline{(s - E) \times r}} + 1 = r^n.$$

Example. — A farmer proposed to a drover that he would sell him 12 sheep and allow him to select them from his flock, provided the drover would pay 1 cent for the first selected, 3 cents for the second, 9 cents for the third, and so on; what sum of money would 12 sheep amount to, at that rate?

$$\frac{r^n \times e - e}{r - 1} = s, \text{ then}$$

$$\frac{3^{12} \times 1 - 1}{3 - 1} = \$2657.20. \quad \textit{Ans.}$$

Note. — Ratio^4, cubed = ratio^{12}; ratio^6, squared = ratio^{12}, &c.

When it is required to find a high power of a ratio, it is convenient to proceed as follows, viz.: write down a few of the lower or leading powers of the ratio, successively as they arise, in a line, one after another, and place their respective indices over them; then will the product of such of those powers as stand under such indices whose sum is equal to the index of the required power, equal the power required.

EXAMPLE. — Required the 11th power of 3.

1	2	3	4	5
3	9	27	81	243

Here $5 + 4 + 2 = 11$, consequently,

$$243 \times 81 \times 9 = 11\text{th power of } 3, \text{ or}$$
$$5 \times 2 + 1 = 11, \text{ consequently},$$
$$243 \times 243 \times 3 = 11\text{th power of } 3, \text{ or}$$
$$4 \times 2 + 3 = 11, \text{ consequently},$$
$$81 \times 81 \times 27 = 11\text{th power of } 3, \text{ or}$$
$$3 \times 3 + 2 = 11, \text{ consequently},$$
$$27^3 \times 9 = r^{11} = 177147. \quad \textit{Ans.}$$

To find any assigned number of geometrical means, between two given numbers or extremes.

RULE. — Divide the greater given number by the less, and from the quotient extract that root whose index is 1 more than the number of means required; that is, if 1 mean be required, extract the square root; if two, the cube root, &c., and the root will be the common ratio of all the terms; which, multiplied by the less given extreme, will give the least mean; and that, multiplied by the said root, will give the next greater mean, and so on, for all the means required. Or the greater extreme may be divided by the common ratio, for the greatest mean; that by the same ratio, for the next less, and so on.

EXAMPLE. — Required to find 5 geometrical means between the numbers 3 and 2187.

$$2187 \div 3 = 729, \text{ and } \sqrt[6]{729} = 3, \text{ then —}$$

$3 \times 3 = 9 \times 3 = 27 \times 3 = 81 \times 3 = 243 \times 3 = 729$, that is, the numbers 9, 27, 81, 243, 729 are the 5 geometrical means between 3 and 2187.

NOTE. — The geometrical mean between any two given numbers is equal to the square root of the product of those numbers. Thus the geometrical mean between 5 and 20, $= \sqrt{(5 \times 20)} = 10$.

ANNUITIES.

An annuity, strictly speaking and practically, is a certain sum of money by the year; payable, usually, either in a single payment yearly, or in half, half-yearly, quarter, quarter-yearly, &c., and for a succession of years, greater or less, or forever. *Pensions, awards, bequests*, and the like, that are made payable in fixed sums for a succession of payments, are commonly rated by the year, and denominated *annuities*.

A current annuity that has already commenced, or that is to commence after an interval of time not greater than that between the stipulated payments, is said to be in *possession*.

One that is to commence or cease on the occurrence of an indeterminate event, as upon the death of an individual, is a *reversionary, contingent*, or *life* annuity.

One that is to commence at a given period, and to continue for a given number of years or payments, is a *certain* annuity.

One that is to continue from a given time, forever, is a *perpetual annuity*, or a *perpetuity*.

Annuity payments do not exist fractionally: they mature, and exist only in that state, and are then due.

A current annuity commences with a payment, and terminates with a payment.

One current in the past is measured from a present included payment, closes with an included payment, and is said to be in *arrears* or *forborne*, from a supposed cancelled payment one regular interval or time beyond.

One current in the future is measured from the present to the first included payment of the series, and from thence is said to *continue* to the close; but if the interval from the present to the first included payment is equal to that between the successive payments, it is supposed to *continue* from the present.

Annuities in negotiation are adjusted, with regard to time, by *interest*, or *discount*, or both.

The TABLES applicable to compound interest and compound discount are applicable in adjusting annuities at compound rates.

To find the Amount of a Current Annuity in Arrears.

LEMMA. — The amount of an annuity that has been forborne for a given time is equal to the sum of the several payments that have become due in that time, plus the interest on each, from the time it became due, until the close of the time.

Then the amount of an annuity of \$100, payable in a single payment annually, but delayed of payment 4 years, allowing simple interest at 6 per cent. on the payments, is

$$\begin{aligned} 100 \times 1.18 &= 118 \\ 100 \times 1.12 &= 112 \\ 100 \times 1.06 &= 106 \\ 100 \times 1 \quad &= 100 = \$436. \end{aligned}$$

And at 6 per cent. compound interest on the payments, it is

$$\begin{aligned} 100 \times (1.06)^3 &= 119.10 \\ 100 \times (1.06)^2 &= 112.36 \\ 100 \times (1.06)^1 &= 106.00 \\ 100 \times \ 1 \quad &= 100.00 = \$437.46. \end{aligned}$$

At 6 per cent. simple interest, when payable in half, half-yearly, it is

$$\begin{aligned} 50 \times 1.21 &= 60.50 \\ 50 \times 1.18 &= 59.00 \\ 50 \times 1.15 &= 57.50 \\ 50 \times 1.12 &= 56.00 \\ 50 \times 1.09 &= 54.50 \\ 50 \times 1.06 &= 53.00 \\ 50 \times 1.03 &= 51.50 \\ 50 \times 1 \quad &= 50.00 = \$442. \end{aligned}$$

And at 6 per cent. compound interest PER ANNUM, when payable in half-yearly instalments, it is

$$\begin{aligned} 50 \times (1.06)^3 \times 1.03 &= 61.34 \\ 50 \times (1.06)^3 &= 59.55 \\ 50 \times (1.06)^2 \times 1.03 &= 57.86 \\ 50 \times (1.06)^2 &= 56.18 \\ 50 \times (1.06)^1 \times 1.03 &= 54.59 \\ 50 \times \ 1.06 &= 53.00 \\ 50 \times \ 1.03 &= 51.50 \\ 50 \times \ 1. &= 50.00 = \$444.02. \end{aligned}$$

From the foregoing, we derive the following general RULES:—

Let $P =$ annuity or yearly sum,
$r =$ rate of interest per annum,
$a =$ rate of discount per annum,
n or $^n =$ nominal time of the annuity in full years,
$A =$ *amount* for the full years,
$D =$ *present worth* for the full years.

When the annuity is payable in a single payment yearly,

$$A = Pn\left(1 + \frac{r(n-1)}{2}\right), \text{ Simple Interest.}$$

$$A = P\,\frac{(1+r)^n - 1}{r}, \text{ Compound Interest.}$$

When payable in equal half-yearly instalments,

$$A = Pn\left(1 + \frac{r(n-1)}{2} + \frac{r}{4}\right), \text{ Simple Interest.}$$

$$A = P \times \frac{(1+r)^n - 1}{r} \times \left(1 + \frac{r}{4}\right), \text{ Compound Interest.}$$

When payable in equal third-yearly instalments,

$$A = Pn\left(1 + \frac{r(n-1)}{2} + \frac{r}{3}\right), \text{ Simple Interest.}$$

$$A = P\,\frac{(1+r)^n - 1}{r}\left(1 + \frac{r}{3}\right), \text{ Compound Interest.}$$

When payable in quarter-yearly instalments,

$$A = Pn\left(1 + \frac{r(n-1)}{2} + \frac{3r}{8}\right), \text{ Simple Interest.}$$

$$A = P\,\frac{(1+r)^n - 1}{r}\left(1 + \frac{3r}{8}\right), \text{ Compound Interest.}$$

When there are odd payments, to find the amount, **S**.

When 1 half-yearly, $S = A(1 + \frac{1}{2}r) + \frac{1}{2}P.$
1 third-yearly, $S = A(1 + \frac{1}{3}r) + \frac{1}{3}P.$
2 " $S = A(1 + \frac{2}{3}r) + \frac{1}{3}P(1 + \frac{1}{3}r) + \frac{1}{3}P$
$= A(1 + \frac{2}{3}r) + P(6 + r) \div 9.$
1 quarter-yearly, $S = A(1 + \frac{1}{4}r) + \frac{1}{4}P.$
2 " $S = A(1 + \frac{1}{2}r) + P(8 + r) \div 16.$
3 " $S = A(1 + \frac{3}{4}r) + P(3 + \frac{3}{4}r) \div 4.$

For any number of equal and regular payments at compound interest per interval between the payments, $S = P'\left(\frac{(1+r')^{n'} - 1}{r'}\right)$, and for any number of equal and regular payments at simple interest per interval between the payments, $S = P'n'\left(1 + \frac{r'(n'-1)}{2}\right)$; P' being a payment, n' or $^{n'}$ the number of payments, and r' the rate of interest per interval between the payments. But this must not be confounded with compound interest *annually*, on payments occurring semi-annually, quarterly, &c.

EXAMPLE. — What is the amount of an annuity of \$150, payable in half, half-yearly, but delayed of payment 2 years and 72 days, allowing compound interest per annum at 7 per cent.?

$150 \times \frac{(1.07)^2 - 1}{.07} = \$310.50$, the amount for 2 years, if payable in yearly payments, and

$310.50 \times (1.\frac{.07}{4}) = \$315.93$, the amount for 2 years, if payable in half-yearly payments, and

$315.93 \times (1.\frac{.07 \times 72}{365}) = \$320.29$, the amount for 2 years and 72 days, if payable in half-yearly payments. *Ans.*

EXAMPLE. — What is the amount of an allowance, pension, or award, of \$100 a year, payable quarterly, but forborne $3\frac{1}{2}$ years, interest compound per annum at 6 per cent. ?

$100 \times \frac{(1.06)^3 - 1}{.06} \times (1 + \frac{.06 \times 3}{8}) = \$325.52$, the amount for 3 years, and

$$325.52\ (1 + .03) + 100 \times 8.06 \div 16 = \$385.66. \quad \textit{Ans.}$$

EXAMPLE. — What is the amount of \$100 a year, payable in quarterly payments, and in arrears 4 years, interest being compound per quarter-year, at 6 per cent. a year ?

$25\left[(1 + \frac{.06}{4})^{16} - 1\right] \times \frac{4}{.06}$. By tabular powers of $(1 + r)$, page 125, $= \$448.30$. *Ans.*

To find the Present Worth of an Annuity Current.

LEMMA. — The present worth of an annuity that is to *continue* for a given time is equal to that sum of money, which, if put at interest from the present time to the close of the payments, will amount to the amount of the payments at that time; and therefore, the times being full, is equal to the sum of the several payments, discounted, respectively, at the rate of interest for their respective times.

NOTE. — If the foregoing proposition is tenable, it follows, since simple interest is due and payable annually, that the true present worth of an annuity having more than one year to run cannot be found by simple interest and discount. By simple interest and discount, at 6 per cent., predicating the rule upon the foregoing lemma, the *amount* of \$100, payable annually, and in arrears for 4 years, is \$436; and the *present worth*, at 6 per cent., is

$$\frac{100}{1.24} + \frac{100}{1.18} + \frac{100}{1.12} + \frac{100}{1.06} = \$340.$$

But \$340 at 6 per cent. interest for 4 years, with the payments of interest annually, will amount to \$440.30; and at interest simply for 4 years it will amount to only \$432.76.

Then the present worth of an annuity of \$100, payable in a single payment yearly, and to continue 4 years, or to become due 1, 2, 3, and 4 years hence, interest and discount being compound per annum, and each at 6 per cent. =

$$\frac{P}{(1+r)^4}+\frac{P}{(1+r)^3}+\frac{P}{(1+r)^2}+\frac{P}{1+r}=\$346.51=$$

$100\times(1.06)^3=119.10$
$100\times(1.06)^2=112.36$
$100\times(1.06)\ \ =106.00$
$100\times\ 1\qquad\ =100.00=437.46\div(1.06)^4=\$346.51.$

And interest at 6 per cent. and discount at 10, both compound, it is

$100\times(1.06)^3=119.10$
$100\times(1.06)^2=112.36$
$100\times\ 1.06\ \ \ =106.00$
$100\times\ 1\qquad\ =100.00=437.46\div(1.10)^4=\$298.79.$

Therefore, when the annuity is payable in a single payment yearly from the present time,

$$D=P\frac{(1+r)^n-1}{r(1+a)^n},=\frac{A}{(1+r)^n}\text{ when } r \text{ and } a \text{ are equal.}$$

When payable in half-yearly payments,

$$D=P\times\frac{(1+r)^n-1}{r(1+a)^n}\times(1+\tfrac{1}{4}r).$$

When payable in third-yearly payments,

$$D=\frac{P\times[(1+r)^n-1]\times(1+\frac{1}{3}r)}{r(1+a)^n}.$$

When payable in quarter-yearly payments,

$$D=\frac{P[(1+r)^n-1](1+\frac{3}{8}r)}{r(1+a)^n}.$$

When there are odd payments, to find the present worth, S.

There being a half-yearly, $S=\frac{D}{1+\frac{1}{2}a}+\frac{\frac{1}{2}P}{1+\frac{1}{2}a}.$

" 1 third-yearly, $S=\frac{D}{1+\frac{1}{3}a}+\frac{\frac{1}{3}P}{1+\frac{1}{3}a}.$

" 2 " $S=\frac{D}{1+\frac{2}{3}a}+\frac{2P(1+\frac{1}{6}r)}{3(1+\frac{2}{3}a)}.$

" 1 quarter-yearly, $S=\frac{D}{1+\frac{1}{4}a}+\frac{\frac{1}{4}P}{1+\frac{1}{4}a}.$

" 2 " $S=\frac{D}{1+\frac{1}{2}a}+\frac{P(1+\frac{1}{8}r)}{2(1+\frac{1}{2}a)}.$

" 3 " $S=\frac{D}{1+\frac{3}{4}a}+\frac{\frac{3}{4}P(1+\frac{3}{8}r)}{1+\frac{3}{4}a}.$

For any number of equal payments, at equal intervals between the payments, $S=P'\times\frac{(1+r')^{n'}-1}{r(1+a')^{n'}}$; P' being a payment, n' the number of payments, and r' and a' the rates per interval between the payments.

NOTE.—Since $\frac{(1+r)^n-1}{r(1+a)^n}$ is the co-efficient of P, for its present worth, at compound interest and discount, for the time n, at the rates r, a, it follows that tables of co-efficients of P for its present worth, at given rates, for any number of years, may be easily made. Thus $(1.06^4-1)\div 1.06^4\times .06=3.46511$, the co-efficient of an annuity, P, for 4 years' continuance, interest and discount being compound per annum, at 6 per cent.; and $(1.06^2-1)\div(1.06^2\times .06)=1.83339$, the co-efficient for 2 years, &c.

If the annuity is deferred, then the difference of two of these co-efficients (one of them that for the time deferred, and the other that for the sum of the time deferred and the time of the annuity) will be the co-efficient of P for its present worth. Thus $3.46511-1.83339=1.63172$, the co-efficient of an annuity, P, for its present worth, when it is to commence two years hence, and to continue 2 years, interest and discount being compound per annum, at 6 per cent. each; or $D=1.63172$ P.

In like manner, tables of other co-efficients, such as the formulæ suggest, may be made that will greatly assist in calculating annuities.

EXAMPLE.—What is the present worth of an award of $500 a year, payable in half-yearly instalments, the 1st payment to mature 6 months hence, and the annuity to continue three years; interest and discount being 7 per cent., compounded yearly?

$$\frac{500\times[(1.07)^3-1]\times\left(1.\frac{.07}{4}\right)}{.07\times(1.07)^3}=\$1335.13.\quad \textit{Ans.}$$

EXAMPLE.—What is the present worth of an annuity of $100, payable in half-yearly payments, and to continue 1½ years; interest and discount being 6 per cent. per annum?

$$D=\frac{100\times[1.06-1]\times 1.\frac{.06}{4}}{.06\times 1.06}=95.755,\text{ and}$$

$$\frac{95.755}{1.03}+\frac{50}{1.03}=\$141.51.\quad \textit{Ans.}$$

EXAMPLE.—What is the present worth of an annuity of $500, payable in semi-annual instalments, and to continue 10½ years, interest and discount being compound per annum, the former at 6 per cent., and the latter at 8?

$$\frac{500\,[(1.06)^{10}-1]\left(1.\frac{.06}{4}\right)}{.06\,(1.08)^{10}\left(1.\frac{.08}{2}\right)}+\frac{500}{2\left(1.\frac{.08}{2}\right)}=$$

$$\frac{A}{1.08^{10}\times 1.04}+\frac{250}{1.04}=\quad \textit{Ans.}$$

By tabular powers of $1 + r$, page 125: —

$\frac{500 \times .79085}{.06 \times 2.15892} = \$3052.64$, the present worth for 10 years' continuance, if payable in yearly payments, and

$$3052.64 \times 1.015 = \$3098.43,$$

the present worth for 10 years' continuance, if payable in half-yearly payments, and

$$3098.43 \div 1.04 + 500 \div 2 \times 1.04 = \$3219.64. \quad \textit{Ans.}$$

When the interval of time from the present to the 1st payment is *shorter* than that between the consecutive payments, and the annuity is payable in a single payment yearly,

$$A = \frac{P[(1+r)^n - 1]\left(1 + \frac{dr}{365}\right)}{r}, \text{ and}$$

$$D = \frac{A}{(1+a)^{(n-1)}\left(1 + \frac{a(365-d)}{365}\right)} = \frac{P[(1+r)^n - 1]\left(1 + \frac{dr}{365}\right)}{r(1+a)^{(n-1)}\left(1 + \frac{a(365-d)}{365}\right)},$$

d being the time in days from the present to the 1st payment.

So, if the annuity is payable in half-yearly, third-yearly, or quarter-yearly instalments, multiply by $1 + \frac{1}{4}r$, $1 + \frac{1}{3}r$, or $1 + \frac{3}{8}r$, as before directed; and if there are odd payments proceed for the present worth, S, as already directed.

EXAMPLE. — Required the present worth of an annuity of $100, payable yearly, to commence 4 months hence, and to continue 4 years; interest and discount being 6 per cent. *annually.*

$$\frac{100 \times (1.06^4 - 1) \times \left(1.\frac{4 \times .06}{12}\right)}{.06 \times 1.06^3 \times \left(1.\frac{.06 \times (12-4)}{12}\right)} = \$360,24. \quad \textit{Ans.}$$

To find the Present Worth of a Deferred Current Annuity, or of an Annuity in Reversion.

When the annuity is payable in a single payment yearly, and the deferred time embraces full years only,

$$D = P\,\frac{(1+r)^n - 1}{r(1+a)^{(n+n')}}, \text{ n' being the deferred time.}$$

If it is payable in half-yearly, third-yearly, or quarter-yearly instalments, multiply by $1 + \frac{1}{4}r$, $1 + \frac{1}{3}r$, or $1 + \frac{3}{8}r$, as already directed; and, if there are odd payments, find the present worth, S, as already directed.

EXAMPLE. — What is the present worth of an annuity of \$150, payable yearly, to commence 2 years hence, and to continue 4 years; interest and discount being compound per annum, at 6 per cent.?

$$150 \times (1.06^4 - 1) \div .06 \times 1.06^6 = \$162.59. \quad \textit{Ans.}$$

EXAMPLE. — Required the present worth of an annuity of \$500, payable in semi-annual instalments, to commence $2\frac{1}{2}$ years hence, and to continue 6 years; allowing compound interest and discount annually at 7 per cent.

$$\frac{500 \times (1.07^6 - 1) \times 1.\frac{.07}{4}}{.07 \times 1.07^8 \times 1.\frac{.07}{2}} = \$2046.44. \quad \textit{Ans.}$$

EXAMPLE. — Required the present worth of an allowance, pension, or award of \$125 a year, payable in half every half-year, to commence 7 months 24 days hence, and to continue $6\frac{1}{2}$ years; interest and discount being compound per annum at 5 per cent.

$$\frac{125 \times (1.05^6 - 1) \times 1.0125}{.05 \times 1.05^6 \times 1.03247 \times 1.025} + \frac{125}{2 \times 1.025} = \$668. \quad \textit{Ans.}$$

Or $\frac{125 \times [(1.05)^6 - 1]}{.05 \times (1.05)^6} = \$634.47$, the present worth for 6 years' continuance, if payable in yearly instalments; and

$$634.47 \times 1.\frac{.05}{4} = \$642.40,$$

the present worth for 6 years' continuance, if payable in half-yearly instalments; and

$$642.40 \div \left(1 + \frac{.05 \times 237}{365}\right) = 622.20,$$

the present worth for 6 years' continuance, if payable in half-yearly instalments, and to commence 7 months, 24 days hence; and

$$622.20 \div (1 + \tfrac{.05}{2}) + \frac{125}{2 \times 1.025} = \$668. \quad \textit{Ans.}$$

To find the Present Worth of a Perpetuity.

LEMMA. — The present worth of an annuity to commence one year hence, and to continue forever, is expressed by that sum of money whose interest for 1 year is equal to the *amount* of the annuity for 1 year; and so, *pro rata*, for perpetuities otherwise regularly affected.

Then when the annuity is to commence 1 year hence, and is payable in a single payment yearly . . . $D = P \div r$.

Payable in half-yearly instalments . . $D = \frac{P(1 + \frac{1}{4}r)}{r}$.

Payable in third-yearly instalments . . $D = \frac{P(1 + \frac{1}{3}r)}{r}$.

Payable in quarter-yearly instalments . $D = \frac{P(1 + \frac{3}{8}r)}{r}$.

EXAMPLE. — What is the present worth of a perpetuity of $150 a year, payable in a single payment yearly from the present time; interest at 6 per cent?

$$150 \div .06 = \$2500. \quad Ans.$$

EXAMPLE. — What is the present worth of a perpetuity of $150 a year, payable in semi-annual instalments, and to commence 4 months hence; interest 7 per cent?

$$\frac{P(1 + \frac{1}{4}r)}{r} + P\,\frac{.07(12 - 4)}{12} = \$2187.36. \quad Ans.$$

EXAMPLE. — Required the present worth of a perpetuity of $400 a year, payable in quarterly payments, and to commence 6 years hence; interest and discount being 5 per cent., compound per year.

$$D = \frac{P(1 + \frac{3}{8}r)}{r(1 + a)^n} = \frac{400 \times 1.\frac{3 \times .05}{8}}{.05 \times 1.05^6} = \$6081.65. \quad Ans.$$

The Amount, Time, and Rate given, to find the Annuity.

When payable in a single payment yearly from the present time,

$$P = \frac{Ar}{(1 + r)^n - 1}; \text{ half-yearly, } P = \frac{Ar}{[(1 + r)^n - 1](1 + \frac{1}{4}r)};$$

$$\text{third-yearly, } P = \frac{Ar}{(1 + \frac{1}{3}r)[(1 + r)^n - 1]}; \text{ quarterly,}$$

$$P = \frac{Ar}{(1 + \frac{3}{8}r)[(1 + r)^n - 1]};$$

and so, *pro rata*, for other fractional units of the integral unit.

Therefore $(1+r)^n - 1 = \frac{Ar}{P}$, or $\frac{Ar}{P(1+\frac{1}{4}r)}$, or $\frac{Ar}{P(1+\frac{1}{3}r)}$

or $\frac{Ar}{P(+\frac{3}{8}r)}$, &c.

Example. — What annuity, payable in quarterly payments from the present time, will amount to \$3000 in 12 years; interest, being compound per annum, at 8 per cent.?

$$3000 \times .08 \div \left[(108^{12} - 1) \times 1.\frac{3 \times .08}{8}\right] = \$153.48. \quad \textit{Ans.}$$

Example. — What length of time must a current annuity of \$400, payable in quarterly payments, remain unpaid, that it may amount to \$2500; interest being 7 per cent. yearly?

$\frac{2500 \times .07}{400 \times 1.\frac{3 \times .07}{8}} = .4263094 = 5 +$ years, and 5 years by table of

$(1+r)^n - 1 = .402552$: therefore $\left(\frac{.4263094}{.402552} - 1\right)\frac{365}{.07} = 308$ days, 5 years, 308 days. *Ans.*

The Present Worth, Time, and Rate given, to find the Annuity.

When payable in a single payment yearly from the present time, $P = \frac{Dr(1+r)^n}{(1+r)^n - 1}$; half-yearly, $P = \frac{Dr(1+r)^n}{[(1+r)^n - 1](1+\frac{1}{4}r)}$; third-yearly, $P = \frac{Dr(1+r)^n}{[(1+r)^n - 1](1+\frac{1}{3}r)}$; quarter-yearly, $P = \frac{Dr(1+r)^n}{[(1+r)^n - 1](1+\frac{3}{8}r)}$, &c. Therefore, $(1+r)^n = \frac{P}{P - Dr} =$

$$\frac{P(1+\frac{1}{4}r)}{P(1+\frac{1}{4}r) - Dr} = \frac{P(1+\frac{3}{8}r)}{P(1+\frac{3}{8}r) - Dr}, \text{ \&c.}$$

Example. — What annuity, payable in half-yearly instalments, and to continue 3 years, is at present worth \$1335.13; discount and interest being compound per year, at 7 per cent?

$$\frac{1335.13 \times .07 \times 1.07^3}{(1.07^3 - 1) \times 1\frac{.07}{.4}} = \$500. \quad \textit{Ans.}$$

OF INSTALMENTS GENERALLY.

Any certain sum of money to be paid on a debt periodically until the debt is paid is called an instalment; and a debt so made payable is said to be payable by instalments.

Let D = principal or debt to be paid,
n = number of years in which the debt is to be paid,
r = rate of interest per annum,
p = instalment or periodical payment.

When the instalments are payable yearly, and the debt is at interest,

$$p = \frac{Dr(1+r)^n}{(1+r)^n - 1}; \quad (1+r)^n = \frac{p}{p - Dr}; \quad D = \frac{p[(1+r)^n - 1]}{r(1+r)^n}.$$

When payable half-yearly,

$$p = \frac{Dr(1+r)^n}{2[(1+r)^n - 1](1 + \frac{1}{4}r]};$$

$$(1+r)^n = \frac{p(1+\frac{1}{2}r)+p}{[p(1+\frac{1}{2}r)+p] - Dr}; \quad D = \frac{2p[(1+r)^n - 1](1+\frac{1}{4}r)}{r(1+r)^n}.$$

When the debt is not on interest, and the instalments are payable yearly,

$$p = \frac{Dr}{(1+r)^n - 1}; \quad (1+r)^n = \frac{Dr + p}{p}; \quad D = \frac{p(1+r)^n - p}{r}.$$

Example. — What yearly instalment will pay a debt of \$4000 in 4 years, the debt being on interest the while, at 6 per cent. annually?

$$4000 \times .06 \times 1.06^4 \div (1.06^4 - 1) = \$1154.37. \quad \textit{Ans.}$$

Example. — What semi-annual instalment will pay a debt of \$1500 in 3 years, the debt bearing interest at 7 per cent. yearly?

$$\frac{4500 \times .07 \times 1.07^3}{2 \times (1.07^3 - 1) \times 1.0175} = \$842.62. \quad \textit{Ans.}$$

When the debt is on interest, and is payable in equal yearly instalments, $p = D(1 + rn) \div n(1 + \frac{r(n-1)}{2})$, at simple interest; but simple interest is not strictly applicable to instalments. See Note, p. 157.

When a debt has been diminished at regular intervals by the payment of a constant sum, to find the remaining debt at the close of the last payment.

When the debt is on interest, and the payments have been made yearly from the date of the debt,

$$d = \frac{p - (p - \mathrm{D}r)(1+r)^n}{r}; \quad (1+r)^n = \frac{p - dr}{p - \mathrm{D}r}$$

$$p = \frac{\mathrm{D}r(1+r)^n - dr}{(1+r)^n - 1}$$

When the payments have been made half-yearly,

$$d = p + p(1 + \tfrac{1}{2}r) - (1+r)^n\,[p + p\,(1 + \tfrac{1}{2}r) - \mathrm{D}r] \div r, \text{ \&c.}$$

EXAMPLE. — On a debt of \$1000, drawing interest the while at 8 per cent. a year, there has been paid yearly, from the date of the debt, \$200 for 6 years: required the unpaid debt at the close of the last payment.

$[200 - 1.08^6(200 - .08 \times 1000)] \div .08 = \$119.69$. *Ans.*

EXAMPLE. — On a note of hand for \$1000, and interest from date, at 8 per cent. annually, the following payments have been made; viz., \$100 at the close of every half-year from the date of the note, for 6 years. How much remained unpaid at the close of the last payment?

$[204 - 1.08^6(204 - .08 \times 1000)] \div .08 = \$90.34$. *Ans.*

NOTE. — In the foregoing, I have treated the terms annual interest and interest payable annually as synonymous in meaning with the terms compound interest and compound interest per annum, and they are so in equity and in fact; besides, simple interest is inapplicable, in equity, to instalment payments. If the debtor stipulates to pay the interest annually on a debt, and abides his contract, he will pay it when it becomes due, and it then becomes a principal in the hands of the creditor, to be let, it is fair to suppose, upon as favorable terms to himself as he let the principal which grew it: whereby he realizes equal to compound interest per annum on the first principal: moreover, if the debtor withholds it from the creditor, it is fair to suppose that he considers it of as much worth to himself as a like part of the principal.

PERMUTATION.

Permutation, in the mathematics, has reference to the greatest number of unlike relative positions, that a given number of things, either wholly unlike, or unlike only in part, may be placed in. It considers the number of changes, therefore, that may be made, in the arrangement of the things, under different given circumstances.

To find the number of changes that can be made in the order of arrangement of a given number of things, when the things are all different.

Rule. — Find the product of the natural series of numbers, from 1 up to the given number of things, inclusive; and that product will be the number of changes or permutations that may be made.

Example. — In how many different relative positions may 12 persons be seated at a table?

$1 \times 2 \times 3 \times 4 \times 5 \times 6 \times 7 \times 8 \times 9 \times 10 \times 11 \times 12 =$ 479,001,600. *Ans.*

To find the number of changes that can be made in the order of arrangement of a given number of things, when that number is composed of several different things, and of several which are alike.

Rule. — Find the number of changes that could be made if the things were all unlike, as in first example. Then find the number of changes that could be made with the several things of each kind, if they were unlike. Lastly, divide the number first found by the product of the numbers last found, and the quotient will be the number of permutations or changes that the collection admits of.

Example. — Required the number of permutations that can be made with the letters a, bb, ccc, $dddd$, = 10 letters.

$$\frac{1 \times 2 \times 3 \times 4 \times 5 \times 6 \times 7 \times 8 \times 9 \times 10 = 3628800}{1 \times 2 \times 6 \times 24 = 288} = 12{,}600. \ \textit{Ans.}$$

To find the number of permutations that can be made with a given number of different things, by taking an assigned number of them at a time.

Rule. — Take a series of numbers beginning with the number of things given, and decreasing by 1 continually, until the number of terms is equal to the number of things that are to be taken at a time; then will the product of the series be the number of changes that may be made.

Example. — What number of changes can be made with the numbers 1, 2, 3, 4, 5, 6, taking three of them at a time?

$6 - 1 = 5,\ 5 - 1 = 4,$ then $6 \times 5 \times 4 = 120$. *Ans.*

What number, by taking 4 of them at a time?

$6 \times 5 \times 4 \times 3 = 360$. *Ans.*

Example. — Arrange the three letters *a*, *b*, *c*, into the greatest number of permutations possible.

abc, *acb*, *bac*, *bca*, *cab*, *cba*, = 6 permutations. *Ans.*

Example. — Arrange the four letters *a*, *b*, *a*, *b*, into the greatest number of permutations possible.

abab, *aabb*, *abba*, *bbaa*, *baba*, *baab*, = 6 permutations. *Ans.*

COMBINATION.

Combination, in the mathematics, has reference to the number of unlike groups, which may be formed from a given number of different things, by taking any assigned number of them, less than the whole at a time. It does not regard the relative positions of the things, one with another, in any of the collections or groups. But it exacts that each group, in all instances, shall have the assigned number of members in it, and that, in every group, in every instance, there shall be a like number of members. It exacts, therefore, that no two groups shall be composed of precisely the same members.

To find the number of combinations that can be made from a given number of different things, by taking any given number of them at a time.

Rule. — Take a series of numbers beginning with that which is equal to the number of things from which the combinations are to be made, and decreasing by 1, continually, until the number of terms is equal to the number of things that are to be taken at a time, and find the product of those numbers or terms. Then take the natural series, 1, 2, 3, 4, &c., up to the number of things that are to be taken at a time, and find the product of that series. Lastly, divide the product first found by the product last found, and the quotient will express the number of combinations that can be made.

Example. — What number of combinations can be made from 8 different things, by taking 4 of them at a time?

$$\frac{8 \times 7 \times 6 \times 5}{1 \times 2 \times 3 \times 4} = \frac{1680}{24} = 70. \quad Ans.$$

What number, by taking 5 of them at a time?

$$\frac{8 \times 7 \times 6 \times 5 \times 4}{1 \times 2 \times 3 \times 4 \times 5} = \frac{6720}{120} = 56. \quad \textit{Ans.}$$

What number, by taking 3 of them at a time?

$$\frac{8 \times 7 \times 6}{1 \times 2 \times 3} = \frac{336}{6} = 56. \quad \textit{Ans.}$$

Example. — What number of combinations can be made from 5 different things, by taking three of them at a time?

$$\frac{5 \times 4 \times 3}{1 \times 2 \times 3} = \frac{60}{6} = 10. \quad \textit{Ans.}$$

What number, by taking 2 of them at a time?

$$\frac{5 \times 4}{1 \times 2} = \frac{20}{2} = 10. \quad \textit{Ans.}$$

Example. — Form 5 letters, a, b, c, d, e, into 10 combinations of 2 letters each; that is, into 10 unlike groups of two letters each.

ab, *ac*, *ad*, *ae*, *bc*, *bd*, *be*, *cd*, *ce*, *de*. *Ans.*

Form them into the greatest number of combinations possible, in collections of three each.

abc, *abd*, *abe*, *acd*, *ace*, *ade*, *bcd*, *bce*, *bde*, *cde*. *Ans.*

SECTION A.

FOREIGN MONEYS OF ACCOUNT, COINS, WEIGHTS, AND MEASURES,

REDUCED TO THEIR VALUES IN THE MONEY, WEIGHTS, AND MEASURES OF THE UNITED STATES.

THE many changes that have been made in the moneys of account, coins, weights, and measures of different countries, by their respective governments, within the last few years, chiefly, though not in all cases, by the adoption of the *Metric System*, or systems bearing aliquot relations thereto, have compelled the author to re-write this section of the work, in a great measure, since the first edition was published; and it is the intention that this edition, and subsequent editions that may be published, shall contain this section strictly correct in all particulars at the time of going to the press.

The Federal units of comparison in the following tables, unless otherwise expressed, are as follows; viz., the *dollar* of 100 cents, in gold; the commercial or avoirdupois *pound*, of 7000 grains; the commercial *yard*, of 36 inches; the commercial or wine *gallon*, of 231 cubic inches; the commercial or Winchester *bushel*, of $2150\frac{42}{100}$ cubic inches; the standard *foot*, of 12 inches; the statute *mile*, statute *acre*, &c.

The value in Federal money, therefore, affixed to any particular denomination of a foreign money of account in the following tables, is the equivalent, or intrinsic par, of that denomination in United-States gold coins. It is predicated upon the standard weight and purity of the coins coined especially to represent that denomination, or conventionally held to be the measure of its value, compared with the standard weight and purity of the gold coins of the United States, that represent the dollar or its multiples.

Thus, in respect to those countries in which gold is made the measure of value and chief legal tender, it is the *intrinsic* par, gold for gold; and, in respect to those countries in which silver is made the measure of value and chief legal tender, it is the par value of that denomination in United-States gold coins, based upon the almost constantly prevailing relative commercial values for many years past, of gold to silver, as 15⅜ to 1, for equal weights. It is, therefore, in a commercial point of view, the ***intrinsic par*** of that denomination, in Federal gold coins, in all cases.

The denomination itself, to which the Federal value is immediately affixed, is usually the integer, or ultimate money of account, of the country especially referred to. It is a money of account in that country always, but not always the name of a circulating coin. Occasionally, even, its value is not represented by any known single circulating coin.

From the foregoing remarks, it will be perceived that, when the *mintage* relative values of gold to silver, in any particular country, are maintained at rates nearer to each other than 15⅜ to 1 for equal weights, the gold coins of that country are commonly worth more, as commercial material, than its silver coins of the same denominations, or same prescribed values; and, conversely, that when the mintage relative values are limited to rates more remote from each other than 15⅜ to 1, the gold coins are commonly worth less than the silver coins.

Thus, the Federal *dollar*, in standard *silver* coins, is ordinarily worth, as commercial material, but $14.88372 \div 15.375 = 96\frac{4}{5}$ cents in Federal gold coins. But, since most Governments make silver the chief measure of value, the *mint* value of gold is usually purposely placed above its commercial worth. Thus, twenty francs in French gold coins are ordinarily worth, as commercial material, but $15.375 \times 20 \div 15.5 = 19.8387$ francs in French silver coins.

It is true that the silver coins of the United States, in small sums, for immediate use, in limited localities at home, may occasionally sell in exchange for the gold coins at their nominal values, or even at a premium, according to the local demand and supply; and the same may happen with regard to the gold coins in exchange for the silver, in France, and those other countries where gold is purposely over-valued in the mintage; but these conditions do not affect the general commercial relations of the metals: they are due only to a slight derangement in the required distributions of the two kinds of coins.

In Germany and Austria, the mint relative values of gold and silver for the Zollverein money, are as 15⅜ to 1, for equal weights.

FOREIGN MONEYS OF ACCOUNT AND COINS REDUCED TO THEIR VALUES IN FEDERAL MONEY.

Foreign.	*U. States.*
ABYSSINIA, (E. AFRICA).—*Massuah:* The old Venetian zechino (*sequin*) is current here at 50 harfs; and 23 harfs = 1 pataka, or old Spanish dollar, - - - - -	= $1.01385
Austrian rix-dollars and Spanish dollars are current here at 1 pataka each.	

NOTE.—The old peso duro colonato, or Carolus silver dollar of Spain, contained, at mint usage, 415 grains of mint silver $\frac{43}{48}$ fine = $1.041353; but it is no longer struck at the mint, and those in circulation are more or less abraded. It is now valued, throughout the British Possessions in North America, and generally, wherever it circulates by tale, or is made the integer of the moneys of account, at 50 pence sterling in gold = $1.0138542. The Austrian rix-dollar (tallaro), scudo, or crown, which, by the way, has not been coined since 1858, except on orders for foreign circulation, contains $\frac{1}{12}$ Vienna mark of fine silver, or 361.11 grains = $1.0114911. This is often called the German dollar; and the Venetian dollar is of the same value.

ALGERIA (N. AFRICA).—*Algiers, Bona, &c.:* 100 centimes = 1 Franc - - - - -	= 0.19452
ARABIA.—*Muscat:* 20 goz = 1 mamooda, 20 m. = 1 current Spanish dollar - - -	= 1.01385

NOTE.—1 goz = 2 paras, and 1 mamooda = 1 piastre of Egypt. See EGYPT.

Mocha, Hodeida: 2 crats = 1 commasse; 60 c. = 1 Mexican or Spanish dollar by tale = 360 grains of fine silver, - - - - - -	= 1.00838
160 crats = 1 wakega or troy ounce (*gold and silver weight*).	
Jidda: Same as at Alexandria, Egypt.	
Aden: 80 caveers = 1 piastre of account = $\frac{5}{6}$ current Spanish dollar, - - - - -	= 0.84488
Also, as at Calcutta. Official, as in Great Britain.	

AUSTRALIA. — *Sidney, Melbourne, Hobart Town, and Australasia generally:*

Standard of purity, denominations, values, and relative values, since 1855, same as in Great Britain.

AUSTRIA. — *Vienna, Prague, Trieste, Ragusa, &c. Zollverein money:* Standard for gold and silver coins = $\frac{9}{10}$ fine, each; relative values, gold to silver as 15.375 to 1.

Foreign.	*U. States.*
4 pfenninge = 1 kreuzer; 60 k. = 1 gulden, or florin = $\frac{1}{45}$ Zollverein pfund (11$\frac{1}{9}$ grams) of fine silver, or 171.471 grains, - - -	= $0.4803
1½ gulden = 1 Zollverein thaler; 1107 gulden = 80 Zoll. krones; 81 gulden = 200 francs; U. S. Customs value of gulden =	
AZORE ISLANDS. — *Fayal, Terceira, Corvo, St. Michael, &c.:*	
1000 reis = 1 milreis of account = $\frac{5}{6}$ old current Spanish dollar, - - - -	= 0.84488
U. S. Customs value = 83$\frac{1}{3}$ cents.	

NOTE. — In 1834, English sovereigns and Spanish dollars were made legal tender here and at the Madeiras; the former at the rate of 4120 reis, and the latter at 870 reis, each, which corresponds very nearly with their intrinsic values in standard Portuguese gold coins.

BALEARIC ISLANDS. — MAJORCA, *Palma;* MINORCA, *Port Mahon:* Same as new system in Spain, see SPAIN.	
BELGIUM. — *Brussels, Antwerp, Ostend, &c.*. Standard for gold and silver coins = $\frac{9}{10}$ fine, each. Relative values, gold to silver as 15.8228* to 1.	
100 centimes = 1 Franc = 4½ grammes of fine silver, - - - - - -	= 0.19452
BERBERA (E. AFRICA.): Same as at Mocha, ARABIA.	
BERMUDA ISLANDS. — Official, as in Great Britain. In trade, 100 cents = 1 dollar = 1 old current Spanish peso, or 50 pence sterling in gold, - - - - - - -	= 1.01385

NOTE. — At the Bermudas, in British America, and other British foreign possessions generally, official or government accounts are kept, and duties to the government are assessed, in sterling money; but until 1842 this class of accounts were kept at the Bermudas and Jamaica, in pounds, shillings, and pence, at 12 shillings sterling to the pound, when it was ordered that hereafter they be kept in sterling money, and that all existing contracts in those colonies be settled at the rate of $\frac{3}{5}$ pound sterling per colonial pound.

BOURBON ISLAND. — *St. Denis:* 100 centimes = 1 franc - - - - - -	= 0.19452

* Although the silver coins of Belgium, both by law and general usage, have the same intrinsic values as those of France of the same denominations, yet this rule does not hold good with regard to the gold coins. The mint standard for 25 francs of France is 8$\frac{2}{81}$ grammes of mint gold $\frac{9}{10}$ fine, while that for 25 francs of Belgium is only 7$\frac{9}{10}$ grammes of mint gold $\frac{9}{10}$ fine; and so in proportion for the other gold coins. 25 francs, French mint, are worth $4.823816, while the 25-franc piece, Belgic mint, is worth only $4.72541; in other words, the Belgic gold coins are less in value two centimes per franc than the French gold coins.

Foreign. — *U. States.*

CANADA, DOMINION OF, *and British America generally:* Standard for silver coins (20-cent pieces or Colonial shillings) = $\frac{37}{40}$ fine; for gold coins (British sovereigns) = $\frac{11}{12}$ fine. Relative values, gold to silver as 14.341 to 1.

100 cents = 1 dollar colonial, - - - = $1.00

Also, 4 farthings = 1 penny; 12 p. = 1 shilling; 20 s. = 1 pound colonial, - - - = 4.00

NOTE.—The standard 20-cent piece, or colonial silver shilling, of British America, contains $66\frac{3}{5}$ grains of fine silver, and is, therefore, worth only 18.655 Federal cents in Federal gold coins, or 19.2708 Federal cents in Federal silver coins. But the money of account shilling in British America is equal to 20 Federal cents in gold; thus, the British pound sterling in gold, the British sovereign, is equal to $4.8665 in gold; and the shillings in that sovereign are equal to 24.3325 Federal cents, each, in gold; therefore, $\frac{4.8665 \times 20}{24.3325}$ = $4.00, the value of the colonial pound in Federal money (gold), measured by, or payable in, British standard gold. See CANADA, page *a* 51.

CANARY ISLANDS. — *Teneriffe, Palma, Grand Canary, Fuerteventura, &c.:* Official as in Spain; in trade, occasionally, 8 reáls (antiquas) of 34 maravedes each = 1 piastre, or peso of exchange, - - - - - = 0.75623

CANDI ISLAND. — Same as in Turkey.

CAPE OF GOOD HOPE (S. AFRICA). — *Cape Town, &c.:* Same as in Great Britain.

CAPE VERDE ISLANDS. — ST. VINCENT, *Mindello;* ST. JAGO, *Porto Praya, &c.:*

1000 reis = 1 milreis. Old Spanish dollars are current here at 870 reis.

Central and South America.

CENTRAL AMERICA. — HONDURAS, *Truxillo, Port San Lorenzo, Omoa, &c.;* NICARAGUA, *Realejo, Greytown, &c.;* SAN SALVADOR, *La Union, Sonsonate, &c.;* COSTA RICA, *Puntas Arenas, Matina, &c.:* GUATEMALA, *Ystapa, &c.:* Standard for silver coins (dollars) = $\frac{20}{171}$ Castilian marco of silver $\frac{171}{200}$ fine, or $\frac{1}{10}$ marco of fine silver to the dollar; for gold coins (double escudos of 32 reáls) = $\frac{5}{177}$ Castilian marco of gold $\frac{171}{200}$ fine. Relative values, gold to silver as 16.5614 to 1.

100 centavos or 8 reáls = 1 dollar = 355.08 grains of fine silver, - - - - = 0.9946

Spanish dollars and U. S. gold coins circulate here, dollar for dollar.

Foreign.	*U. States.*
BALIZE, *Balize:* Official accounts are kept here in sterling money.	
SOUTH AMERICA.—PERU, *Callao, Islay, Truxillo, Arica, &c.;* CHILI, *Valparaiso, Concepcion, Coquimbo, &c.;* NEW GRANADA, *Cartagena, Santa Martha, Savanillo, Buenaventura, &c.;* ECUADOR, *Guayaquil, &c.* The prescribed standard for the mintage of these States is now in conformity with that of France, and there is strong probability that Brazil, and the other States in South America having mints, will soon adopt the same standard. Standard for gold and silver coins = $\frac{9}{10}$ fine, each. Relative values, gold to silver as $15\frac{1}{2}$ to 1.	
100 cents = 1 sol, or dollar = $22\frac{1}{2}$ grammes of fine silver, - - - - - - =	$0.97461
Also, 100 centesimas = 1 duro or old Spanish dollar, - - - - - - =	1.01385
BRAZIL.—*Rio Janeiro, Maranham, Bahia, Para, Pernambuco, &c.:* Standard for silver coins = $\frac{1}{18}$ Castilian marco of silver $\frac{11}{12}$ fine per milreis; for gold coins = $\frac{4}{103}$ Castilian marco of gold $\frac{11}{12}$ fine per 10 milreis. Relative values, gold to silver as 14.30556 to 1.	
1000 reis = 1 milreis = 180.8278 grains fine silver, - - - - - - =	0.50651
A current Spanish dollar passes for 2 milreis of account, and the modern gold coins (10 milreis and 20 milreis) = $0.544375 per milreis.	
BOLIVIA.—*Cobija, &c.:* Standard for silver coins (dollars) = $\frac{4}{23}$ marco of silver $\frac{11}{12}$ fine; for gold coins (doubloons) = $\frac{2}{17}$ marco of gold $\frac{7}{8}$ fine. But little gold is coined in Bolivia, and that in circulation has, at present, no nominal mint relation to the silver coins.	
100 centavos = 1 dollar = 283.03478 grains of fine silver, - - - - - =	0.79292

NOTE.—The gold coins of Bolivia are often light of weight, and seldom range above $\frac{87}{100}$ fine. The silver dollars are usually minted at full weight, but are often but little if any above $\frac{9}{10}$ fine. The fractional silver coins are worth less, relatively, than the integer.

VENEZUELA.—*La Guayra, Maracaybo, Cumana, Puerto Cabello, &c.:* In Venezuela, as in Peru, &c.,

Foreign. | *U. States.*

the silver 5-franc piece of France is made the measuring unit of value, and is divided into 100 centavos. In the moneys of account, however, the peso macuquins of 80 centavos is sometimes used; and one dollar in United-States gold coin is assumed to be worth 7 cents more than 5 silver francs, which is the case at the metal ratio of gold to silver as 16.0794 to 1, for equal weights. Hence, 1 peso fuerte Americano de premio = $\frac{107}{80}$ = $1.3375, measured by the peso macuquins.

ARGENTINE REPUBLIC. — *Buenos Ayres, Parana, &c.;* URUGUAY, *Montevideo, &c.;* PARAGUAY, *Assumption, Neembucu, &c.:*

Foreign gold and silver coins circulate here measured generally by the Spanish dollar: 10 decimos = 1 reál; 8 r. = 1 peso.

GUIANA, *Cayenne:* Same as in France.

Paramaribo: Same as in Holland.

Georgetown: 100 cents = 1 dollar = 50*d.* sterling.

FALKLAND ISLANDS. — Same as in Great Britain.

CHINA. — *Canton, Shanghai, Amoy, Ningpo, Foochoo, Hongkong I., Macao:*

Standard for gold and silver ingots = 94 touch, or $\frac{94}{100}$ fine, each. Gold is not treated as money, but as merchandise.

10 cash or lc = 1 candarine or fun; 10 c. = 1 mace or tsëen; 10 m. = 1 tael or leäng = 583$\frac{1}{3}$ grains of silver $\frac{94}{100}$ fine, or 548$\frac{1}{3}$ grains of fine silver, - - - - - = $1.53591

NOTE. — Mexican dollars, which are about the only coins, except copper cash, that circulate in China, on account of their convenience generally, bear a premium well laid on, upon their intrinsic worth. They commonly pass for about 72 candarines each, or $\frac{72}{100}$ of the tael in ingots of standard purity. A tael weight, or leäng as it is called by the natives, of wan-yin or sysce (see-sze) silver, that is, of silver as fine as silk, or that may be drawn into a thread as fine as silk (pure silver), is valued by the East India Company at 80 pence sterling. In London, it is commonly valued by the price per ounce paid at the mint for the fine silver in foreign silver coins, which is about 66 pence sterling, or $\frac{66 \times 583\frac{1}{3}}{480}$ = 80.2 pence sterling per tael. In 1799, the Congress of the United States fixed the value of the tael, in ingots of standard purity, for customs' purposes, at $1.48; but this was when silver was the measure of value in the United States, and when the dollar contained 371$\frac{1}{4}$ grains of fine metal. The law, it is believed, has not been repealed.

CORSICA ISLAND. — *Ajaccio:* 100 centimes = 1 franc, - - - - - - = 0.19452

Foreign.	*U. States*
CYPRUS ISLAND. — Same as in Turkey.	
DENMARK. — *Copenhagen, Elsinore, Odense, &c.:*	
Standard for silver coins = $\frac{877}{1000}$ fine.	
16 skillinge = 1 mark; 6 m. 1 rigsdaler = $\frac{75}{1216}$ Cologne mark of mint silver = $\frac{1}{2}$ former speciesdaler = 195.1233 grains fine silver, -	= $0.546552
U. S. Customs value of speciesdaler = $1.05.	

NOTE. — Denmark coins ducats and 5-thaler and 10-thaler gold pieces; but these and the speciesdaler have no proposed mint-relation to each other. The ducat is of the ordinary value of that denomination of coins; and the 10-thaler piece is designedly and practically of the same intrinsic value as that of North Germany; viz., equal to ten Bremen thalers in silver at the metal ratio of 15.409 to 1; or equal, at the common par of exchange, to $7.90005 in gold.

EGYPT (N. AFRICA). — *Cairo, Alexandria, Suez, &c.:* 40 paras = 1 piastre; in current gold coins = $0.04996647; in current silver coins = $0.05054979.

NOTE. — In 1836, British sovereigns were made legal tender throughout Egypt at 97 pi., 20 pa., each; Spanish doubloons or onzas at 313 pi., 29 pa.,; Napoleons (20 francs in gold) at 77 pi., 6 pa.; Venetian sequins at 46 pi., 13 pa.; Dutch ducats at 45 pi., 26 pa.; tallaros (German dollars) at 20 pi.; colonatos (Spanish dollars) at 20 pi., 28 pa.; 5-franc pieces at 19 pi., 10 pa. The measure of value is $\frac{8}{8}$ oka-drachmas of fine silver to the piastre = $2\frac{2}{9}$ Abyssinian dirhem or 18 troy grains = $0.050419109.

FRANCE. — Standard for gold and silver coins = $\frac{9}{10}$ fine, each. Relative values, gold to silver as $15\frac{1}{2}$ to 1.	
10 centimes = 1 decim; 10 d. = 1 franc = $4\frac{1}{2}$ grammes of fine silver, - - -	= 0.19452
In practice, 100 centimes = 1 franc. U. S. Customs value = $0.186.	
GERMANY (*The Zollverein States, or North German Confederation*): PRUSSIA, SAXONY, MECKLENBURG, OLDENBURG, *the* HANSE CITIES, *&c. Zollverein money:* Standard for gold and silver coins = $\frac{9}{10}$ fine, each. Relative values, gold to silver as 15.375 to 1.	
12 pfennige = 1 silber groschen; 30 s. g. = 1 thaler = $\frac{1}{30}$ Zollverein pfund ($16\frac{2}{3}$ grams) of fine silver, or 257.2058 grains, - -	= 0.72045
369 Zollverein thalers = 40 Zollverein krones; 1 Zoll. krone = $\frac{1}{50}$ Zollverein pfund (10 grams) of fine gold, or 154.3235 grains = $6.64614; 27 thalers = 100 francs.	
U. S. Customs value of thaler =	

NOTE. — Saxony reckons 10 pfennige to the groschen.

Foreign.	*U. States.*
Bremen (special): 5 schwaren = 1 groot; 72 g. = 1 thaler = $\frac{1}{5}$ old Frederic d'or, -	=$0.79619
U. S. Customs value = $78\frac{3}{4}$ cents.	
Hamburg, Lubec, Altona (special): 12 pfennige = 1 schilling; 16 s. = 1 mark.	
1 marc current = $\frac{1}{35}$ Cologne mark of fine silver,	= 0.28869
1 marc banco, at par (London rate of 13*m.* 12*s.* to the £.), - - - - - -	= 0.35393
U. S. Customs value of mark current = 28 cents; of mark banco = 35 cents.	
Southern States.—BAVARIA, WURTEMBURG, BADEN, *&c.*: Standard for silver coins = $\frac{9}{10}$ fine.	
4 pfennige = 1 kreuzer; 60 k. = 1 gulden or florin = $\frac{2}{105}$ Zoll. pfund of fine silver ($9\frac{11}{21}$ grams) or 146.975 grains, - - -	= 0.41169
7 gulden = 4 Zollverein thalers; 189 gulden = 400 francs.	
U. S. Customs value of gulden = 40 cents.	
GREAT BRITAIN.—*Sterling money:* Standard for silver coins = $\frac{37}{40}$ fine; for gold coins = $\frac{11}{12}$ fine. Relative values, gold to silver as 14.287891 to 1.	
4 farthings = 1 penny; 12 p. 1 shilling; 20 s. = 1 pound, - - - - -	= 4.86656
U. S. Customs value = $4.84.	

NOTE.—Since 1816, the Government of Great Britain has estimated the value of gold compared with that of silver as $14\frac{1393}{4840}$ to 1, for equal weights. In theory, by the mint regulations, a *pound sterling* is equal to $1614\frac{6}{11}$ grains of fine silver, or $113\frac{1}{623}$ grains of fine gold. A pound sterling in silver, therefore, measured by the Federal standard of $345\frac{3}{5}$ grains of fine silver to the dollar, is equal to $\$4\frac{133}{198}$ in Federal silver coins; or a standard silver shilling, measured by this measure, is equal to $23\frac{71}{198}$ cents; and at the ordinary par of exchange it is worth $0.226122, very nearly, in Federal gold coin. But a pound sterling in British gold (a sovereign) measured by the Federal standard of $23\frac{11}{50}$ grains of fine gold to the dollar, is equal to $\$4\frac{626788}{723303}$, or $4.866563529, very nearly, in Federal gold coins. At the former relative values of the two currencies, or relative values before the Federal standard was changed, in 1834, viz., 4 *shillings* and 6 *pence sterling* to the *dollar*, or $\$4.44\frac{4}{9}$ to the *pound*, the par of exchange is .09498, or $9\frac{1}{2}$ per cent, practically, in favor of sterling money. Gold is the measure of value in Great Britain, as well as in the United States; and the silver coins of that government are not legal tender at home in sums exceeding £2.

GREECE.—*Athens, Patras, the Ionian Islands, &c.*: Standard for gold and silver coins = $\frac{9}{10}$ fine, each. Relative values, gold to silver as 15.549 to 1.	
100 lepta = 1 dramia = 1 dramia weight of fine silver, or $62\frac{1}{8}$ grains, - - - -	= 0.17401

Foreign.	*U. States.*
5 dramias = 1 taliron. 20 dramias (gold) = 1 othonion = $3.44137.	
100 oboli =1 Spanish, German, or Venetian dollar = $1.01385.	

NOTE. — Greece is a party to the present movement (1869) to legalize and enforce the use of the metric system of moneys, weights, and measures, throughout Continental Europe.

HOLLAND (NETHERLANDS). — *Amsterdam, Rotterdam, The Hague, &c.:*

Standard for silver coins = $\frac{113}{125}$ fine; for gold coins = $\frac{9}{10}$ fine. Relative values, gold to silver as 15.7333 to 1.	
100 centimes = 1 guilder or florin = 10 grams of mint silver, - - - - -	= $0.40806
Gouden Willem (10 guilders) = 6$\frac{2}{3}$ grams of mint gold = $3.98769.	
U. S. Customs value of guilder = 40 cents.	

HAWAIIAN OR SANDWICH ISLANDS. — *Honolulu, &c.:*

100 cents = 1 dollar, - - - -	= 1.00

India and Malaysia, or East Indies.

HINDOSTAN. — BENGAL, MADRAS, BOMBAY, *Presidencies of:* Standard for gold and silver coins = $\frac{11}{12}$ fine, each. Relative values, gold to silver as 15 to 1.

Calcutta, Madras, Rangoon, &c.: 12 pice = 1 anna; 16 a. = 1 rupee = 1 tola, or 180 grains of mint silver, = $\frac{15}{16}$ old sicca rupee, - - -	= 0.46217
1 mohur, or gold rupee (E. I. Co.), = 15 silver rupees in theory = 1 tola of mint gold, or 165 grains of fine gold, = $7.10594.	
Bombay, Surat, &c.: 100 reas = 1 quarter; 4 q. = 1 rupee, - - - - -	= 0.46217
U. S. Customs value of rupee = 44$\frac{1}{2}$ cents.	
Madras, &c. (old usage): 48 jittas = 1 fanam; 36 f. = 1 star pagoda = 4 arcot, or Company rupees, - - - - - -	= 1.84869
48 jittas = 1 fanam; 36 f. = 1 India pagoda = 8 shillings sterling in gold, - - -	= 1.9466
U. S. Customs value of star pagoda = $1.84; of Indian pagoda, = $1.94.	
Goa: Official, same as in Portugal.	
Pondicherry: Same as in France.	

Foreign.	*U. States.*
CEYLON ISLAND. — *Colombo, Trincomalee, &c.:* Official, same as in Great Britain; also, 1 rixdaalder = 18*d.*; 1 Spanish dollar = 50*d.*; 1 rupee = 22*d.*, sterling.	
MALAYA. — *Malacca, Pahang, Perak, &c.:* Same as at Singapore.	
PENANG ISLAND. — Same as in Calcutta; also, 100 cents = 1 Spanish dollar.	
SIAM. — *Bangkok, &c.:* 4 prangs, or clams, = 1 sompay; 4 s. = 1 salung; 4 salungs = 1 tical = 1 tical weight of silver 93 touch, or $\frac{93}{100}$ fine, - =	$0.61659
U. S. Customs value of tical = 61 cents.	
SINGAPORE ISLAND. — *Singapore:* 100 cents = 1 dollar (old Spanish), - - - - =	1.01385
BANCA ISLAND. — Same as at Batavia (JAVA I.).	
BORNEO ISLAND. — *Banjermassin, Sarawak, &c.:* Generally as at Batavia.	
CELEBES ISLAND. — *Macassar*, and the other Dutch settlements: Same as at Batavia.	
JAVA ISLAND. — *Batavia, Samarang, &c.:* 100 cents = 1 guilder, - - - - - =	0.40806
Mexican dollars are current here at $2\frac{3}{8}$ guilders, each.	
MOLUCCA ISLANDS. — *Amboyna, &c.:* Same as at Batavia (JAVA I.).	
PHILIPPINE ISLANDS. — LUZON, *Manila, Mindano, &c.:* Official, as in Spain; also, 34 maravedis = 1 reál; 8 r. = 1 peso, or dollar, - - =	1.00465
SUMATRA ISLAND. — *Bencoolen, Padang:* Official, as at Batavia; also, 8 satellers = 1 soocoo; 4 s. = 1 dollar, or rial, = 1 pardow of Acheen.	
Acheen: 16 copangs = 1 mace; 10 mace of fine silver, or 374.306 troy grains = 1 pardow, - =	1.04845
Mexican and Spanish dollars pass for 1 pardow, each.	

ITALY. — The recent formation of the several States in Italy into one kingdom, Rome and in its immediate vicinity only excepted, has had the effect to unify the moneys and moneys of account in that portion of Europe also; and now, throughout Italy proper, or the Kingdom of Italy, including the Island of Sicily and that of Sardinia, the official monetary system, with a slight difference in nomenclature, is identically the same as that of France.

Standard for gold and silver coins = $\frac{9}{10}$ fine, each.

Relative values, gold to silver as $15\frac{1}{2}$ to 1.

Foreign. *U. States.*

100 centesimi = 1 lira, or franco, = $4\frac{1}{2}$ grams of fine silver, (See SICILY, page *a* 51.) - = $0.19452

JAPAN. — *Yeddo, Miaco, Osaka, Simoda, Hakodadi, Nagasaki, Yokohama, Matsmay, Napa, &c.: New System:* Standard for silver coins (assay) = $\frac{89}{100}$ fine; for gold coins = $\frac{143}{250}$ fine. Relative values, gold to silver (assuming the cobang to represent 10 silver itzebous) as 15.01877 to 1.

50 sen = 1 itzebou, or itchibou (often called *boo*) = 5 monme of mint silver, or $\frac{89}{20}$ monme of fine silver = 119.1888 grains, - - - = 0.333855

1 cobang contains $\frac{518}{100}$ monme of mint gold, or 79.35992 grains of fine gold, - - - = 3.417744

U. S. Customs value of itzebou =

NOTE. — Mexican dollars circulate in Japan, commonly at 2.87 to 2.90 itzebou, each; but they are intrinsically worth over 3 itzebous, each, by tale.

LIBERIA (W. AFRICA). — *Monrovia, &c.:* Same as in the United States.

MADEIRA ISLANDS. — *Funchal, &c.:* 1000 reis = 1 milreis of account, - - - = 1.01385

See Azore Islands, note relative to.

U. S. Customs value of milreis = $1.00.

MALTA ISLAND. — *Valetta, &c.:* Official, as in Great Britain; also, 20 grani = 1 taro; 12 t. = 1 scudo = $\frac{1}{2}$ ducat of Naples, - - = 0.41371

U. S. Customs value of scudo = 40 cents.

MAURITIUS ISLAND. — *Port Louis, &c.:* Official, as in Great Britain; also, 100 cents = 1 dollar. Tallaros of Austria and silver Napoleons (5-franc pieces of France) are current here at 1 dollar each; Spanish pesos and Mexican dollars, at 52 pence sterling each.

MEXICO. — *Acapulco, Mazatlan, San Blas, Campeachy, Tampico, Sisal, Vera Cruz, &c.:*

Standard for silver coins (dollars, average by assay) = 416.15 grains $\frac{9}{10}$ fine; for gold coins (doubloons, average by assay) = 416.4 grains $\frac{433}{500}$ fine. Relative values, gold to silver as 16.618192 to 1.

100 cents = 1 dollar; also, 6 grani = 1 cuarto; 2 c. = 1 medio; 2 m. = 1 reál; 8 r. = 1 dollar = 374.535 grains of fine silver, - - = 1.0491

NOTE. — The Mexican marco = 3549.81 grains; and the standard silver dollar should weigh $\frac{3549.81 \times 40}{341}$ = 416.4 grains.

Foreign.	*U. States.*
MOROCCO (N. AFRICA). — *Morocco, Fez, Tangier, &c.*: 24 fluce = 1 blankeel; 10 b. = 1 metical; 4 m. = 1 oncia or ducat = 4 meticals weight of fine silver, or 295.3846 grains = 1 old standard ducat of Naples, - - - -	= $0.82739
MOZAMBIQUE (E. AFRICA). — *Mozambique, Quilimane, Sofala, Delagoa Bay, &c.*: 1000 reis = 1 milreis of account.	

NOTE. — A variety of foreign coins are current here, and many of them wide of their true values; viz., Spanish and Mexican dollars at 1000 reis, each; silver 5-franc pieces of France and United-States gold coins per dollar at 900 reis, each; Spanish doubloons at 17120 reis, patriot doubloons at 16000 reis; and British sovereigns at 4500 reis, each.

NORWAY. — *Christiania, Bergen, &c.*: Standard for silver coins = $\frac{877}{1000}$ fine.	
20 skillinge = 1 mark; 6 m. = 1 speciesdaler = 2 rixdalers = $\frac{75}{608}$ Cologne mark of mint silver or 390.2465 grains of fine silver, - -	= 1.0931

NOTE. — Norway and Sweden are about to coin gold of 10 and 20 francs.

NUBIA (E. AFRICA). — *Suakin, &c.*: Same as at Alexandria, Egypt.	
PERSIA. — *Bushire, Gombroon, Astrabad, &c.*: 100 mamoodis, or 50 abasse, = 1 toman = $\frac{3}{4}$ miscal weight of fine gold, or 49.23077 grains, -	= 2.12019
5 dinars = 1 kasbequis; 2 k. = 1 dinars-biste; 2 d.-b. = 1 shatree, shafree, or shahis; 2s. = 1 mamoodi; 2 m. = 1 abasse; 2½ a. = 1 papabat; 2 p. = 1 saheb-keran.	

NOTE. — Foreign gold and silver coins in great variety circulate in Persia, but at fluctuating prices; the toman, however, is commonly valued at two Mexican dollars.

PORTUGAL. — *Lisbon, Oporto, St. Ubes, &c.*: Standard for gold coins (coroas) = $\frac{1}{24}$ marco of gold $\frac{11}{12}$ fine; for silver coins (crusados) = $\frac{1}{16}$ marco of silver $\frac{9}{10}$ fine. Relative values, gold to silver as 14.72524 to 1.	
1000 reis = 1 milreis = 2 crusados = 398.4237 grains of fine silver, - - - -	= 1.11601
1 milreis in gold = $1.16525.	
Method of writing and reading monetary quantities: *Example.* — Rs. 5 : 600 ⊕ 750 = 5,600 milreis and 750 reis.	
U. S. Customs value of milreis = $1.12.	

NOTE. — In commercial transactions, lately, Mexican dollars commonly pass for 1 milreis each, and United-States gold coins at 970 reis per dollar.

2

Foreign.	*U. States.*
ROME *and Civita Vecchia:* Standard for gold and silver coins = $\frac{9}{10}$ fine, each. Relative values, gold to silver as $\frac{24.2\times125}{196}$ = 15.43367 to 1.	
5 quatrini = 1 baiocho; 10 b. = 1 paolo; 10 p. = 1 scudo, or crown, = $24\frac{1}{5}$ grams of fine silver	= $1.04609
The gold coins are $2\frac{1}{2}$, 5, and 10 scudo pieces, and contain $\frac{196}{125}$ grams of fine gold per scudo.	
U. S. Customs value of scudo = $1.05.	
RUSSIA. — *St. Petersburg, Riga, Cronstadt, Odessa, &c.:* Standard for silver coins = $\frac{7}{8}$ fine; for gold coins (assay) = $\frac{229}{250}$ fine. Relative values, gold to silver as 15.17295 to 1.	
10 kopecks = 1 grieven; 10 g. = 1 rublyu (*rouble*) = $\frac{20}{451}$ funt of fine silver, or 280.19 grains	= 0.78483
In practice, 100 kopecks = 1 rouble.	
100 silver roubles = 360 bank or paper roubles. Bank rouble varies from par to 4 per cent premium.	
U. S. Customs value of silver rouble = 75 cents.	
SENEGAMBIA (W. AFRICA). — *Gambia, Bathurst, Sierre Leone, &c.:* Official, as in Great Britain.	
St. Louis: Official, as in France.	
SOCOTRA ISLAND. — Same as at Muscat, Arabia.	
SPAIN. — *Madrid, Malaga, Cadiz, Santandre, Bilboa, Barcelona, &c.: New system,* legalized Oct. 19, 1868, and its use made obligatory to the exclusion of all other systems after Dec. 31, 1870. Standard for gold coins = $\frac{9}{10}$ fine; for silver coins (5-peseta pieces, duros, or pesos) = $\frac{9}{10}$ fine; 1-peseta pieces and less = $\frac{167}{200}$ fine. Relative values, gold to silver (duros) as $15\frac{1}{2}$ to 1.	
100 centesimas = 1 peseta = $4\frac{1}{2}$ gramos of fine silver = 1 franc of France, - - -	= 0.19452

NOTE. — In this system the peseta of account is equal to 1 silver franc of France; and the 5-peseta silver coin is equal to 5 francs in silver of France; but the 1-peseta coin is worth but $0.18047. The gold coins are worth peseta for franc in gold.

Prevailing system last preceding the foregoing: Standard for silver coins (escudos of 10 reáls vellón) = $\frac{9}{10}$ fine; for gold coins (10 escudos and over) = $\frac{112}{125}$ fine. Relative values, gold to silver as 15.5555 to 1.	
1000 milesimas = 1 escudo = $\frac{5}{99}$ marco of fine silver, or $179\frac{1}{3}$ grains, - - - -	= 0.50232

NOTE.—The silver coins of less denominations than 4 reáls, and the gold coins of less denominations than 10 escudos, have less purity than those mentioned above; the 80-reáls piece, even, is but $\frac{1789}{2000}$ fine = \$3.85446, while the 100-reáls piece (double de Isabel) is worth \$4.96493.

Foreign.	*U. States.*
Gibraltar: 16 quartos = 1 reál; 12 r. = 1 peso duro or old Spanish dollar, - - -	= \$1.01385
SWEDEN.—*Stockholm, Gothenburg, Carlscrona, Gefle, &c.:* Standard for silver coins = $\frac{3}{4}$ fine; for gold coins (ducats) = $\frac{39}{40}$ fine. Relative values, gold to silver as 15.18517 to 1.	
100 öre = 1 riksmynt, or riksdollar riksgäld = $\frac{1}{4}$ speciesdollar, - - - - -	= 0.27622
1 speciesdollar = $\frac{15}{124}$ mark of fine silver, or 393.15 grains, - - - - -	= 1.10124
U. S. Customs value of speciesdollar = \$1.06.	
1 ducat = 8 riksmynts, nominally, = $\frac{3}{205}$ marks of fine gold = \$2.23738.	
Sweden is about to coin gold coins of 10 and 20 francs.	
SWITZERLAND.—*Basel, Bern, Geneva, Lausanne, Lucerne, Neufchatel, Zurich, &c.:*	
Standard for gold and silver coins = $\frac{9}{10}$ fine, each. Relative values, gold to silver as $15\frac{1}{2}$ to 1.	
10 rappen = 1 batzen; 10 b., or 100 centimes, = 1 franc = $4\frac{1}{2}$ grams of fine silver, - -	= 0.19452
TRIPOLI (N. AFRICA).—*Tripoli, &c.:* 100 paras = 1 piastre, or ghersch; value of piastre same as that of Tunis.	
TUNIS (N. AFRICA).—*Tunis, Soosa, Cabes, &c.:*	
Standard for silver coins = $\frac{11}{12}$ fine; for gold coins = $\frac{9}{10}$ fine. Relative values, gold to silver as 15.8125 to 1.	
2 burbine = 1 asper; 52 a., or 16 karob = 1 piastre = $\frac{1}{12}$ drachma of fine silver, or 44 grains,	= 0.123247
TURKEY.—*Constantinople, Smyrna, Aleppo, Trebizonde, &c.:* Standard for silver coins (assay) = $\frac{83}{100}$ fine; for gold coins (assay) = $\frac{183}{200}$ fine. Relative values, gold to silver as 15.15625 to 1.	
40 paras, or 100 aspers, = 1 piastre = $\frac{1}{320}$ checki of fine silver, or 15.3856 grains, - -	= 0.043096
1 piastre in gold = $\frac{1}{4850}$ checki of fine gold = \$0.0437181. 100 piastres in gold = 1 medjdie.	
1 purse of silver = 500 piastres; 1 purse of gold = 30,000 piastres. U. S. Customs value of piastre = 5 cents.	

West Indies.

CUBA ISLAND. — *Havana, Matanzas, Santiago, Manzanillo, Baracoa, Cardenas, Cienfuegos, Nuevitas, Trinidad, &c:* Official as in Spain; also, 12 dineros, or 16 quartos, = 1 reál; 8 r., or 100 centesimas = 1 duro, peso, piastre, or dollar.

NOTE. — The full weight Spanish duro colonato is made the unit, or measure of value, = $1.04872; but the Spanish onza, or doubloon, passes for 17 dollars, and the Mexican and S. American, for 16 dollars.

HAYTI ISLAND. — HAYTI, *Port au Prince, Aux Cayes, Cape Haytien, Gonaives, &c.:*

DOMINICA, *San Domingo, Porto Plate, Samana, &c.:* 100 centesimas = 1 dollar, or gourda (old Spanish).

NOTE. — Spanish doubloons pass for 16 dollars of account; and the Haytian silver gourda is worth about ⅛ Spanish dollar.

PORTO RICO ISLAND. — *San Juan, Guayama, Ponce, &c.:* Official, as in Spain; also, 100 centesimas = 1 dollar.

JAMAICA ISLAND. — *Kingston, Falmouth, Savana la Mar, &c.:* Official, as in Great Britain; also, 100 cents = 1 dollar = 1 old peso, or duro of Spain = 50 pence sterling in gold, - - = $1.01385

CARIBBEE ISLANDS. *LEEWARD ISLANDS.* — ANTIGUA, — *St. John, Falmouth;* DOMINICA, MONTSERRAT, TORTOLA, VIRGIN GORDA, ST. CHRISTOPHER, ANGUILLA, BARBUDA, NEVIS, SABA: Same as in Jamaica.

ST. EUSTACIUS: Official, as in Holland.

GUADELOUPE, ST. MARTIN,* MARIE GALANTE, DESIRADE, LES SAINTES: Official, as in France; also, 100 centimes = 1 dollar; the colonial *livre* of these islands = $\frac{1}{10}$ standard duro colonato of Spain.

ST. THOMAS, — *Charlotte Amalie;* SANTA CRUZ, ST. JAN: Official, as in Denmark; also, 100 cents = 1 dollar.

ST. BARTHOLOMEW: Official, as in Sweden; also, 1 Spanish dollar = 9 shillings currency.

* The southern portion of the island of St. Martin is owned and settled by the Dutch; and the moneys of account, weights, and measures of Holland are in general use there.

Foreign. *U. States.*

WINDWARD ISLANDS. — TRINIDAD, *Port Spain*; BARBADOES, *Bridgetown*; GRENADA, ST. VINCENT, ST. LUCIA. TOBAGO: Same as at Jamaica Island.

MARTINIQUE, *St. Pierre, Port Royal*: Official, as in France.

BAHAMA ISLANDS. — NEW PROVIDENCE, *Nassau*; TURKS, ABACO, ANDROS, GREAT BAHAMA, &c.: Same as at Jamaica Island.

LITTLE ANTILLE ISLANDS. — CURACOA, BUEN AYRE, ORUBA: Official, as in Holland.

MARGARITA, TORTUGA, BLANQUILLA: Same as in Venezuela.

ZANGUEBAR (E. AFRICA). — *Zanzibar* (Island and Town), *Quiloa, Mombas, Magadoxo, Brava, Socotra Island, &c.*: Accounts are now kept in Zanzibar and the Sultan of Muscat's dominions on the east coast of Africa generally in *dollars* of 100 *cents*; and by the exertions of William E. Hines, Esq., of New York, late resident consul of the United States at Zanzibar, the present standard dollar in gold of the United States is made the measure of value, or, in other words, is officially rated at par. This was accomplished in 1865. Other foreign gold and silver coins circulate at conventional rates, some of them above and others below their intrinsic values. Thus, the Spanish dollar = $1; the Austrian rix-dollar, scudo, or crown, sells for $1.01 to $1.03; the English sovereign is rated at 4⅗ Austrian rix-dollars; the French franc in gold, at 18¾ cents; the silver 5-franc piece, at 94 cents; and the Indian silver rupee, at 47 cents.

FOREIGN LINEAL AND SURFACE MEASURES,

REDUCED TO THE LINEAL AND SURFACE MEASURES OF THE UNITED STATES.

Foreign.	*U. States.*	
ABYSSINIA. — *Massuah:* 8 robi = 1 derah, or pic, - - - - - =	0.682	yard.
ALGERIA. — 10 decimetres = 1 metre, - =	1.0936	"
8 robi = 1 pic. Pic, *Moorish*, for linens, - =	0.519	"
Pic, *Turkish*, for silks, &c., - =	0.692	"
ARABIA. — *Muscat:* 8 gheria = 1 covid; 8 c. = 1 kassaba, - - - - =	$12.86\frac{2}{3}$	feet.
500 kassaba = 1 coss, - - - =	1.2185	miles.
Aden: 8 robi = 1 yard or pic, - - =	0.95	yard.
Jidda: 8 robi = 1 pic, - - - =	0.743	"
Mocha: 8 robi = 1 gez, - - - =	0.694	"
AUSTRALIA. — Same as in Great Britain.		
AUSTRIA. — *Imperial and Legal:* 12 zollen = 1 fuss, - - - - - =	1.04	feet.
$29\frac{1}{2}$ zollen = 1 elle, - - - =	0.85216	yard.
6 fusse = 1 klafter; 10 fusse = 1 ruthe, - =	10.39924	feet.
2,400 ruthen = 1 meile, - - - =	4.7269	miles.
192 square ruthen = 1 metzen; 3 m. = 1 joch, - - - - - =	1.43	acres.
AZORE ISLANDS. — Same as in Portugal.		
BALEARIC ISLANDS. — 3 pie, or 4 palmi, = 1 vara.		
MAJORCA: 2 varas = 1 cana, - - =	1.711	yards.
MINORCA: 2 varas = 1 cana, - - =	1.754	"
BELGIUM. — 10 streep = 1 duim; 10 d. = 1 palm; 10 p. = 1 el = 1 metre, - - =	1.0936	"
10 elen = 1 roed; 100 r. = 1 mijl, *or* kilometre, - - - - =	0.6214	mile.
100 square roeds = 1 bunder = 100 ares - =	2.471	acres.
The old Brabant el = 0.76006 yard.		
BERMUDA ISLANDS. — Same as in Great Britain.		
BOURBON ISLAND. — Same as in France.		
CANARY ISLANDS. — Same as in Spain.		
CANDIA ISLAND. — 8 rob = 1 pic, - - =	0.697	yard.
CAPE VERDE ISLANDS. — Same as in Portugal.		

Foreign.		*U. States.*	
CANADA, DOMINION OF. — Same as in the United States.			
LOWER CANADA (*special*): 1 arpent,	- =	0.8475	acre.
CAROLINE ISLANDS. — Same as in Spain.			

Central and South America.

CENTRAL AMERICA. — COSTA RICA, GAUTEMALA, HONDURAS, NICARAGUA, SAN SALVADOR: Same as Spain, — *standard of Castile.*			
SOUTH AMERICA. — ARGENTINE REPUBLIC: 36 pulgadas = 3 pes = 1 vara,	- =	34.1	inches.
150 varas = 1 cuadra; 40 c. = 1 legua,	- =	3.229	miles.
27,000 square varas = 1 suertes des estancia.			
BRAZIL: 24 pollegadas = 2 pes = 1 covado = $\frac{66}{100}$ metre, - - -	- =	0.7218	yard.
40 pollegadas = 5 palma = 1 vara = ½ bracio = $1\frac{1}{10}$ metre, - -	- =	1.203	"
250 varas, *or* 125 braccios = 1 estadio; 8 e. = 1 milha; 3 m. = 1 legoa, -	- =	4.101	miles.
4,840 square varas = 1 geira, -	- =	1.447	acres.
CHILI: 36 pulgadas = 1 vara (*customs*),	- =	1.000	yard.
In all other respects as in Spain, — *Castilian standard.*			
VENEZUELA, NEW GRENADA, ECUADOR, PERU, BOLIVIA, PARAGUAY, URUGUAY: Same as Spain, — *Castilian standard.*			
GUIANA: *Cayenne:* Same as in France.			
Paramaribo: Same as in Holland.			
Georgetown: Same as in Great Britain.			
FALKLAND ISLANDS: Same as in Great Britain.			
CHINA. — 10 tsun, or fan, = 1 punt; 10 p. = 1 chih, covid, *or* cobre (*mercers'*), -	- =	1.226	feet.
17½ punts, *or* 10 tac, = 1 thuoc (*tradesmen's*)	=	0.7152	yard.
Chih (*mathematical*), - -	- =	13.122	inches.
Chih (*engineers' and surveyors'*), -	- =	12.059	"
Chih, *or* kong-pu (*architects'*), -	- =	12.709	"
10 chih = 1 chang *or* cheung; 10 chang = 1 yan; 18 y. = 1 li, - -	- =	0.3425	mile.
100 square chih = 1 mow; 240 m. = 1 fu, *or* king, - - - - -	- =	0.5564	acre.
CYPRUS ISLAND. — 8 robi = 1 pic,	- =	0.696	yard.

Foreign.		*U. States.*	
DENMARK. — 24 tomme, *or* 2 fod, = 1 aln, -	=	0.6862	yard.
10 fod = 1 rode; 2,400 r. = 1 mül, -	=	4.6785	miles.
280 square rodes = 1 skiepper; 2 s. = 1 toende, - - - - -	=	1.362	acres.
6 fod = 1 favn; and 1 fod = 1 Prussian Rhein-fuss.			
EGYPT. — 3 kirat = 1 rob; 8 r. = 1 pic *or* endrasi (*for silks and woolens*), - -	=	27.06	inches.
1 halebi *or* archin = 27.9 in.; 1 derah = 25.264 inches.			
2 derah = 1 fedan; 3 f. = 1 gasab; 420 g. = 1 berri; 3 b. = 1 parasang, - -	=	3.015	miles.
400 square gasab = 1 fadden al risach, -	=	1.465	acres.
FRANCE. — 100 centimètres, *or* 10 decimètres, = 1 mètre, - - - -	=	39.37	inches.
100 mètres, *or* 10 decamètres, = 1 hectomètre; 100 hectomètres, *or* 10 kilomètres, = 1 myriamètre, - - -	=	6.2137	miles.
100 square metres = 1 are; 100 a. = 1 hectare, - - - - -	=	2.471	acres.
1 old aune de Paris = 1.29972 yard; 1 aune *usuelle*, or *metrique*, = 1⅕ metre.			
FRIENDLY ISLANDS (FEEJEE AND TONGA GROUPS). — Same as in England.			
GERMANY. — PRUSSIA, and the Zollverein längenmaasse for all Germany:			
12 linie = 1 zoll; 12 z. = 1 fuss (Rheinfuss), - - - - -	=	12.3514	inches.
25⅓ zollen (Rhein-zollen) = 1 elle = ⅔ metre, - - - - -	=	26.24⅔	"
10 land-fusse, *or* 12 Rhein-fusse, = 1 ruthe; 2,000 r. = 1 meile, - - -	=	4.67855	miles.
180 square ruthen = 1 morgen, - -	=	0.6304	acre.
Special and local: —			
SAXONY: 12 zollen = 1 fuss; 2 f. = 1 elle, -	=	0.6196	yard.
1 Lubec-Brabant elle, - - -	=	0.7498	"
BAVARIA: 120 zollen, *or* 10 fusse, = 1 ruthe,	=	9.579	feet.
34¼ zollen = 1 elle = ⅚ metre, - -	=	32.808⅓	inches.
WURTEMBURG: 100 zollen, *or* 10 fusse = 1 ruthe, - - - - -	=	9.37326	feet.
21⅓ zollen = 1 elle, - - -	=	0.67175	yard.
BADEN: 20 zollen, *or* 2 fusse, = 1 elle = ⅗ metre, - - - - -	=	23.622	inches.

Foreign.	*U. States.*
HESSE DARMSTADT: 10 zollen = 1 fuss = ¼ metre, - - - - -	= 0.822 $\frac{1}{12}$ foot.
24 zollen = 1 elle = ⅝ metre, - -	= 0.6561⅜ yard.
MECKLENBURG: Same as at Hamburg.	
OLDENBURG: 12 zollen = 1 fuss; 2 f. = 1 elle,	= 0.63528 "
BREMEN: 24 zollen, *or* 2 fusse, = 1 elle, -	= 0.63276 "
1 Bremen-Brabant elle = 1⅛ Bremen elle.	
HAMBURG: 12 zollen = 1 fuss; 2 f. = 1 elle,	= 0.62681 "
1 Hamburg-Brabant elle, - - -	= 0.75615 "
LUBEC: 24 zollen, *or* 2 fusse = 1 elle, -	= 0.6294 "
GREAT BRITAIN.— Same as in the United States.	
GREECE.— 60 onué = 5 pes = 1 passo = 1¾ metre.	
22 onué = 1 pichi, *for silks*, - -	= 0.701734 "
23½ onué = 1 pichi, *for woolens, &c.*, -	= 0.749579 "
HOLLAND.— 10 streep = 1 duim; 10 d. = 1 palm; 10 p. = 1 el = 1 metre, -	= 39.37 inches.
10 el = 1 roed; 100 roed = 1 mijl = 1 kilometre.	
100 square roede = 1 bunder = 1 hectare.	
1 old Brabant el - - - -	= 0.75931 yard.

India and Malaysia, or East Indies.

ANAM: Same as in China.	
BURMAH, PEGU: 5½ pulgaut = 1 taim; 4 t. = 1 sadang; 7 s. = 1 sha, *or* bambou, -	= 154 inches.
1,000 dhas = 1 dain, - - -	= 2.4306 miles.
CEYLON ISLAND: 2 covids = 1 guz, *or* yard, -	= 36 inches.
HINDOSTAN.— *Calcutta:* 16 tussoos = 8 gheira = 1 haut, *or* covid; 2 h. = 1 ghes, *or* guz,	= 36 "
2,000 ghes = 1 coss, - - -	= 1.1364 miles.
20 square covids = 1 chattack; 16 c. = 1 cottah; 20 cottahs = 1 biggah, - -	= 0.3306 acre.
Bombay: 16 tussoos = 8 gheira = 1 haut *or* covid; 1½ h. = 1 guz; 2 hauts, *or* 1⅓ guz, = 1 imperial yard, - - -	= 36 inches.
Madras: 16 tussoos = 8 gheira = 1 haut *or* covid; 2 h. = 1 guz, - - -	= 36 "
Goa: Same as in Portugal.	
Massulipatam: 2 palms = 1 span; 3 s. = 1 cubit *or* covid, - - - -	= 19⅛ "
Surat: 18 tussoos = 1 haut, *for matting*, -	= 21 "
84 tussoos, *or* 20 wiswūsa, = 1 wūsa, -	= 98 "

Foreign.	*U. States.*
Malabar Coast. — *Mangalore, Cananore, Calicut, Cochin, Quilon, Trivandrum:* 3 gheria = 1 ady; 2½ a. = 1 haut; 2 h. = 1 guz *or* gujah, - - - -	= 38½ inches.
Coromandel Coast, generally, same as at Madras.	
Siam. — 48 nions = 4 keubs = 2 soks = 1 ken,	= 37.836
2 kens = 1 vouah; 20 v. = 1 sen; 4 s. = 1 jod; 25 j. = 1 roëneng, - -	= 2.3886 miles.
Malacca. — Same as at Calcutta.	
Singapore Island. — Same as at Calcutta.	
Philippine Islands. — Same as in Spain, — *standard of Castile.*	
Java Island. — 1 voot (Rhyn-voot), - -	= 1.029 feet.
1 el, *or* covid (old Amsterdam el), - -	= 0.7522 yard.
English measures also, as at Sumatra.	
Sumatra Island. — 2 tempohs = 1 jancal; 2 j. = 1 etto; 2 e. = 1 hailoh, - -	= 36 inches.
ITALY. — The metric system of weights and measures is now the official standard, and is in general commercial use throughout the kingdom.	
Special and local: —	
Milan: 10 atomi = 1 dito; 10 d. = 1 palmo; 10 p. = 1 metro *or* braccia, - -	= 39.37 inches.
1,000 metri = 1 miglio (kilometre).	
100 metri quadrata = 1 tavolo (are); 100 t. = 1 tornatura (hectare).	
2½ metri = 1 trabucco; 4 t. = 1 decametre.	
Florence, Leghorn, Pisa, &c.: 2 palmi = 1 braccia; 4 b. = 1 canna = 2⅓ metres.	
Carrara: 1 palm for marble = 9.06 U. S. inches.	
JAPAN. — 10 rin = 1 bun; 10 b. = 1 tsun; 10 t. = 1 sasi *or* sjak.	
4 tsun-sasi = 5 kani-sasi = 1 hiro; 5 tsun-sasi = 1 ink *or* tattamy, - - -	= 2.07103 yards.
7 kani-sasi = 1 kan, kian, *or* ikje, - -	= 2.31955 "
60 inks = 1 ting *or* masti; 36 t. = 1 ri, -	= 2.54172 miles.
MADEIRA ISLANDS. — Same as in Portugal.	
MAURITIUS ISLAND. — Official as in Great Britain; also, —	
1 aune (old aune de Paris), - -	= 46.79 inches.
MEXICO. — Same as in Spain, *standard of Castile.*	

Foreign.	*U. States.*	
MOROCCO. — 8 pollegadas = 1 palmo da craveira (Portuguese).		
18$\frac{2}{3}$ pollegadas, *or* 2$\frac{1}{3}$ palmi da craveira, = 1 covado *or* cadee, - - -	= 20.21	inches.
8 rob = 1 pic (Turkish in theory), -	= 26.03	"
NORWAY. — Same as in Denmark.		
NEW ZEALAND ISLANDS. — Same as in England.		
PERSIA. — *Monkelser or royal:* 12 fingers = 1 foot; 1$\frac{1}{2}$ f. = 1 cubit; 2 c. = 1 guz *or* gueza (*for silks, &c.*), - - -	= 36.92	"
1 guz shah (royal), *for woolens*, - -	= 40	"
1 guz tabree (of Tabreez), - -	= 40.4	"
6,000 guz (monkelser) = 1 parasang, -	= 3.496	miles.
PORTUGAL. — 8 pollegadas = 1 palmo da craveira = $\frac{22}{100}$ metre.		
12 pollegadas = 1 pé; 2 p. = 1 covado = $\frac{66}{100}$ metre.		
40 pollegadas = 1 vara = 1$\frac{1}{10}$ metre, -	= 43.307	inches.
24$\frac{3}{4}$ pollegadas = 1 covado avantejado (retail).		
13$\frac{1}{3}$ pollegadas = $\frac{1}{3}$ vara = 1 terça.		
60 pollegadas = 1 passo; 1$\frac{1}{3}$ p. = 1 braça.		
780 pés = 1 estado; 8 e. = 1 milha; 3 m. = 1 legua, - - - -	= 3.8386	miles.
4,840 square varas = 1 geira, - -	= 1.4471	acres.
ROME. — 12 onze = 1 palmo; 1$\frac{1}{3}$ p. = 1 pié; 5 piédi = 1 passo, - - -	= 58.6104	inches.
10 palmi, *or* 7$\frac{1}{2}$ piédi, = 1 canna architectona, - - - - -	= 87.9156	"
3$\frac{3}{4}$ canne arch. = 1 catena (*chain*); 1 catena = 10 stajoli.		
8 palmi mercantile = 1 canna merc., -	= 78.43932	"
9 palmi d'are = 1 canna d'are = 1$\frac{1}{8}$ metre,	= 44.29$\frac{1}{8}$	"
1,000 passi = 1 miglio, - - -	= 0.92504	mile.
3$\frac{1}{2}$ square catene = 1 quartuccio; 2 q. = 1 scorzo; 4 s. = 1 quarta; 4 quarte (7 pezze) = 1 rubbio, - - -	= 4.559	acres.
RUSSIA. — 6$\frac{6}{7}$ verschok = 1 foot, - -	= 12	inches.
16 verschok = 1 archine *or* halebi, -	= 28	"
3 archines (7 feet) = 1 sachine, - -	= 7	feet.
500 sachines = 1 verst, - - -	= 0.6629	mile.
2,400 square sachines = 1 deciatene, -	= 2.7	acres.
SANDWICH ISLANDS. — Same as in the United States.		

Foreign.		*U. States.*
SOCIETY ISLANDS. — Official, as in France.		
SPAIN. — *Standard of Castile:* 36 pulgadas = 6 sesma = 4 quarta *or* palmos = 3 tercia *or* pie = 1 vara, - - -	- =	33.385 inches.
2 varas = 1 estado, braza, brazada, *or* toesa.		
2 estados = 1 estadale; 2,000 estadales = 1 legua, - - - -	- =	4.2153 miles.
560 square estadales = 1 fanegada,	- =	1.592 acres.
$1\frac{1}{2}$ pie = 1 codo; 5 pies = 1 passo.		
Gibraltar: As above; also, 12 inches = 1 foot; 3 f. = 1 yard, - - -	- =	1 yard.
Alicante: 36 pulgadas = 1 vara, -	- =	0.8319 "
Barcelona: 1 vara *or* matja-cãna, -	- =	0.8641 "
Santander: 8 octava, *or* 4 palma, = 1 vara,	- =	0.9142 "
Valencia: 36 pulgadas = 4 palmos = 3 pie = 1 vara, - - -	- =	1.0044 yards.
$2\frac{1}{4}$ varas = 1 braza-reale.		

NOTE. — In 1849, Spain legalized the use of the metric system of weights and measures in all her dominions; and now, since she is to enforce the employment of the money system of France, to the exclusion of all others (see Foreign Moneys of Account, Spain), it may be expected that she will unify her weights and measures by enforcing the employment of the metric system generally.

SWEDEN. — 10 linier = 1 tum; 10 t. = 1 fot; 10 f. = 1 stäng; 10 s. 1 ref, -	- =	97.406 feet.
20 tumen = 2 fus = 1 aln = $\frac{1}{2}$ old aune de Paris, - - - -	- =	0.64937 yard.
6 fus = 1 famn; 6,000 f. (3,600 stänger) = 1 mil, - - - -	- =	6.6413 miles.
560 square stänger (14,000 square alner) = 28 kannland = 16 kappland = 2 spannland = 1 tunnland, - -	- =	1.21975 acres.
SWITZERLAND. — Official and general for the 22 cantons forming the Republic; —		
10 zollen = 1 fuss; 2 f. = 1 elle = $\frac{6}{10}$ metre,	=	23.622 inches.
4 fusse = 2 ellen = 1 stab = 1 aune metric.		
6 fusse = 1 klafter; 10 fusse = 1 ruthe *or* toise, - - - -	- =	9.8425 feet.
1,600 ruthen = 1 meile *or* hour's way = $4\frac{4}{5}$ kilometres, - - - -	- =	2.9826 miles.
400 square ruthen = 1 juchart *or* feld-aker = 36 ares, - - -	- =	0.8896 acre.
TRIPOLI. — 8 rob = 3 palmi = 1 pic *or* dra,	- =	26.42 inches.
The pic for silks = 1 Arabic covid, -	- =	19.03 "

Foreign.		U. States.	
TUNIS. — 8 robi = 1 pic; pic for woolens,	- =	26.5	inches.
Pic for silks, - - - -	- =	24.83	"
Pic for linens, - - -	- =	18.62	"
TURKEY. — *Constantinople and Smyrna:* 1 halebi *or* archin (Russian archine of 28 U. S. inches in theory), - - -	- =	27.9	"
7,500 halebi = 1 agatch, - -	- =	3.3142	miles.
1 pic, draa, *or* endrasi, *for silks and woolens,*	=	27.06	inches.
1 indise, *or* endese, *for cottons, &c.,* -	- =	25.688	"
Bassorah, Bagdad: 8 robi = 1 guz, -	- =	31⅜	"

West Indies.

In the Islands of *Cuba, Hayti, Porto Rico,* and *Isle of Pines,* the measures of length and of surface are the same as in Spain, *Castilian standard,* except that in Port au Prince and the French portion of Hayti, generally, the old piede du roy of 12.78918 U. S. inches, and the old aune de Paris of 1.29972 U. S. yards, is used.

In *Jamaica, St. Kitts, Antigua, Montserrat, Tortola, Anguilla, Dominica, Barbuda, Nevis, Virgin Gorda;* in *Trinidad,* Grenada, St. Vincent, St. Lucia, Barbadoes, Tobago, Grenadines;* and in *New Providence, Great Bahama, Turks, Abaco,* and the Bahamas generally, the lineal and surface measures are the same as in Great Britain and the United States.

In *Gaudeloupe, Martinique, Desirade, Les Saintes, Marie Galante,* the lineal and surface measures are the metric; but the pied américaine (United-States foot) is chiefly used in measuring timber.

In *Santa Cruz, St. Thomas, St. John,* the official measures of length and of surface are the same as in Denmark, but those of the United States are much used.

In *St. Bartholomew:* Official, as in Sweden; also as in the United States.

In *St. Eustatius, Curacoa, Buen Ayer, Oruba:* Official, as in Holland.

In *Margarita, Blanquilla, Tortuga:* Official, as in Venezuela.

ZANGUEBAR (East Africa): Same as at *Muscat,* Arabia.

* The Spanish (Castilian) vara is still used to some extent by the merchants in Trinidad.

FOREIGN COMMERCIAL WEIGHTS,

REDUCED TO THEIR EQUIVALENT VALUES IN THE UNITED STATES.

Foreign.	*U. States.* Avoirdupois pounds.
ABYSSINIA (E. AFRICA).—*Massuah:* 10 dirhem = 1 wakea; 12 w. *or* 10 mocha = 1 rotl, rotolo *or* litre = 10 troy ounces,* - - =	0.68571
ALGERIA (Barbary, N. AFRICA).—*Algiers, Bona, Oron:* The metric weights, under French denominations, are in official and common use here.	
ARABIA.—*Mocha:* 233⅓ krat = 1 wakea, wakega, *or* vacia = 700 troy grains.	
15 wakea = 1 rotl; 2 r. = 1 maund *or* maon, - =	3.—
10 maunds = 1 frazil; 15 f. = 1 bahar, - =	450.—
At the bazaar, 14½ frazils = 1 bahar for coffee, =	435.—
Jidda: Official, as at Alexandria, Egypt; also, 3⅝ drachmas musr (of Egypt) = 1 wakea; 15 w. = 1 rotl; 5 r. = 1 maund; 10 m. = 1 frazil; 10 f. = 1 bahar = 66⅔ okes of Egypt, - =	182.8571
Hodeida, Beit-el-fakih; 15 wakea = 1 rotl; 2 r. = 1 maund; 10 m. = 1 frazil; 40 f. = 1 bahar, =	815.2381
16½ rotl = 1 tomaun of rice, - - - =	168.14286
Muscat, Hasek, &c.: 233⅓ krat = 1 wakea; 10 w. = 1 rotl; 9 r. = 1 maund; 200 m. = 1 bahar, =	1800.—
Aden: Official, same as in Great Britain.	
AUSTRALASIA (OCEANICA). — AUSTRALIA, NEW ZEALAND, TASMANIA: Same as in Great Britain.	
AUSTRIA (*legal for the Empire*): 32 lothe = 16 unzen = 4 vierdinge = 2 marken = 1 pfund; 20 p. = 1 stein; 5 s. = 1 centner, - - =	123.468
4 centners = 1 karch; 5 k. = 1 last; 2¾ centners = 1 saum; 1¼ centners = 1 lagel; 2 l. = 1 saum for steel.	
Ragusa, (DALMATIA): 2½ pfunde = 1 oka, - =	3.0867
AZORE ISLANDS (N. Atlantic Ocean): Same as in Portugal.	

* See Turkey, weights of, and note relative to.

Foreign.	*U. States.* Avoirdupois pounds.
BALEARIC ISLANDS (Mediterranean Sea): 25 rotoli (22⅓ Castilian libras) = 1 aroba; 4 a. = 1 quintal, - - - - - =	91.3063
3 quintals = 1 carga; 110 rotoli = 1 oder; 100 libra menor = 87 Castilian libra = 1 cantaro grosso = 88.2627441 av. lbs.	
BELGIUM. — Same as in France (metric system).	
BERBERA (E. AFRICA): Same as at Mocha, Arabia.	
BERMUDA ISLANDS (N. Atlantic Ocean): Same as in Great Britain.	
BOURBON ISLAND (Mascerene group, Indian Ocean): The metric system, French nomenclature, is in use here.	
CANADA, DOMINION OF. — Same as in the United States.	
CANARY ISLANDS (Atlantic Ocean, W. coast N. Africa): Same as in Spain, Castilian standard.	
CANDIA ISLAND, OR CRETE (E. Mediterranean Sea): 100 rottoli = 44 okes = 1 cantaro, - - - - - - =	116.565
CAPE OF GOOD HOPE (S. AFRICA): Same as in Great Britain.	
CAPE VERDE ISLANDS (Atlantic Ocean, near W. African coast): Same as in Portugal.	

Central and South America.

GUATEMALA, HONDURAS, SAN SALVADOR, NICARAGUA, COSTA RICA: 16 onzas = 2 marcos = 1 libra; 25 l. = 1 arroba; 4 a. = 1 quintal, - - - - - =	101.4514
2½ quintals = 1 carga; 8 c. = 1 tonelada, - =	2029.0285
BALIZE. — Official, same as in Great Britain.	
BRAZIL. — 16 onças = 2 marcos = 1 arratel, - =	1.01187
32 arratels = 1 arroba; 4 a. = 1 quintal, - =	129.5193
ARGENTINE REPUBLIC, *or* LA PLATA: 16 onças = 2 marco = 1 libra; 25 l. = 1 arroba; 4 a. = 1 quintal, - - - - =	101.274

NOTE. — The Argentine Republic has established independent standards of weights and measures, which are now in practice, and which vary more or less in each department from those of Castile.

Foreign.	*U. States.* Avoirdupois pounds.
PERU, CHILI, BOLIVIA, ECUADOR, NEW GRANADA, VENEZUELA, URUGUAY, PARAGUAY: Generally as in Central America.	
Montevideo: 1 pesado of dry hides (fresh or salted) contains $1\frac{1}{2}$ arrobas; 1 pesado of wet salt hides contains $2\frac{1}{3}$ arrobas.	

NOTE.—In Peru, Chili, New Granada, Bolivia, Venezuela, and Surinam, the use of the metric system of weights and measures is sanctioned by law; but as yet (1869) is very little employed in either of the States.

GUIANA.—*Cayenne:* Same as in France.	
Paramaribo, or *Surinam:* Same as in Holland; also as in France.	
Georgetown: Same as in Great Britain.	
FALKLAND ISLANDS.—Same as in Great Britain.	
CHINA.—10 tsëen = 1 tael *or* leäng; 16 t. = 1 catty *or* kan; 100 c. = 1 pecul *or* tam, -	= 133.3333
$22\frac{3}{4}$ chu = 1 leäng; 2 catties = 1 yin; 15 y. = 1 kwan; $3\frac{1}{3}$ k. = 1 tam; $1\frac{1}{5}$ t. (60 yin) = 1 shik, - - - - - -	= 160.—
CORSICA ISLAND (Mediterranean Sea): Same as in France, of which it forms a department.	
CYPRUS ISLAND (Mediterranean Sea): Same as in Turkey, and forming a part of Turkey in Asia.	
DENMARK.—32 lod = 16 unze = 2 marken = 1 pund = $\frac{1}{2}$ kilogram; 100 p. = 1 centner, -	= 110.2311
16 pund = 1 lispund; 20 l. = 1 shifpund; $16\frac{1}{4}$ s. = 1 last.	
12 punds = 1 bismerpund; 3 b. = 1 waag *or* vog.	
EGYPT (N. AFRICA): 4 gran = 1 kara; 16 k. = 1 drachma (oka-drachma) = $\frac{1}{10}$ troy ounce, or 48 grains.	
400 drachmas = 4 oka, - - -	= 2.74286
144 drachmas = 1 rottolo; 100 r. (36 okes) = 1 cantaro (customs), - - - -	= 98.7429
FRANCE.—1000 milligrammes = 100 centigrammes = 10 decigrammes = 1 gramme = 15.43235 troy grains.	
1000 grammes = 100 decagrammes = 10 hectogrammes = 1 kilogramme, - -	= 2.20462
10 kilogrammes = 1 myriagramme; 100 m. = 10 quintals = 1 tonneau, - - -	= 2204.62143

Foreign.	*U. States.* Avoirdupois pounds.
GERMANY. — Zollverein gewighte for all the States of the tariff-alliance:	
10 quentchen = 1 loth; 30 l. = 1 pfund; 100 p. = 1 centner = 50 kilograms, - -	= 110.2311
Special and local, or domestic: —	
512 pfennige = 128 quentchene = 32 lothe = 16 unzen = 2 marken = 1 pfund.	
PRUSSIA: 100 pfunde = 1 centner, - -	= 103.1194
BAVARIA: 100 pfunde = 1 centner = 56 kilograms,	= 123.4588
BREMEN: 116 pfunde = 1 centner, - -	= 127.4887
BRUNSWICK: 100 pfunde = 1 centner, - -	= 103.0656
HAMBURG: 112 pfunde = 1 centner, - -	= 119.6044
HESSE DARMSTADT: 100 pfunde = 1 centner = 50 kilos, - - - - -	= 110.2311
LUBEC: 112 pfunde = 1 centner, - - -	= 119.6813
MECKLENBURG: 14 pfunde = 1 liespfund; 8 l. = 1 centner, - - - - -	= 119.5164
OLDENBURG: 100 pfunde = 1 centner; 3 c. = 1 pfundschwer, - - - -	= 317.7241
10 pfunde = 1 liespfund; 29 l. = 1 schiffpfund.	
WURTEMBERG: 100 pfunde = 1 centner, . -	= 103.1153
BADEN: 10,000 as = 1000 dekas = 100 centas = 10 zehnling = 1 pfund.	
100 p. = 10 stein = 1 centner = 50 kilograms, -	= 110.2311
SAXONY: 10,000 as = 1000 dekas = 100 hektas = 10 kilas = 1 pfund; 100 p. = 10 halbstein (half-stones) = 1 centner = 50 kilos, -	= 110.2311
Leipsic (domestic): 100 pfund = 1 centner, -	= 103.0734
GREAT BRITAIN. — Same as in the United States.	

NOTE. — In Great Britain, in addition to the denominations of weights used in the United States (the values of which are the same), the

Clove of wool, - - -	= 7 lbs.	Stone of butchers' meat or flesh,	= 8 lbs.
Stone " " iron, flour, -	= 14 "	Stone of cheese, - - - -	= 16 "
Tod " " - - -	= 28 "	Stone of glass, - - - -	= 5 "
Weigh " " - - -	= 182 "	Seam of glass, - - - -	= 120 "
Sack " " - - -	= 364 "	Stone of hemp, - - - -	= 32 "
Last " " - - -	= 4368 "	Fother of lead, - - - -	= 19½ cwt.

GREECE. — 72 cocos = 1 dramia; 8 d. = 1 ounghia; 8 o. = 1 imilitron; 2 i. = 1 litra, -	= 1.136
400 dramias = 3⅛ litras = 1 oka, - -	= 3.55
137½ litras = 44 okas = 1 cantaro, - -	= 156.2
HOLLAND. — 10,000 korrel = 1000 wigtje = 100 lood = 10 onz = 1 pfund = 1 kilogram, -	= 2.20462

Foreign.	*U. States.* Avoirdupois pounds.
HAWAIIAN ISLANDS (Sandwich Islands, Polynesia, N. Pacific Ocean): Same as in the United States.	

India and Malaysia, or East Indies.

ANNAM. — *Kesho* (Tonquin): 100 catties = 1 pecul,	= 132.—
Hue, Sai-gon, &c. (Cochin China): 16 leäng = 1 can; 10 c. = 1 yen; 5 y. = 1 binh; 2 b. = 1 ta; 5 t. = 1 quan, - - - -	= 688.76
BURMAH (Farther India), *Prome, Patanago, Ava, &c.:* 100 tical *or* kiat = 8 abucco = 4 agito = 3 catties = 1 vis *or* visay, - - -	= 3.39286
CEYLON ISLAND (Indian Ocean): 500 pond = 1 bahar *or* candy, - - -	= 500.—
HINDOSTAN. — *Bombay:* 72 tanks = 30 pice = 1 maund, - - - - -	= 28.—
Also, 80 tipprees = 40 seers = 1 maund, -	= 28.—
20 maunds = 1 candy = 5 cwt.	
Calcutta, Bengal (*factory weight*): 5 siccas *or* rupees = 1 chattac; 16 c. = 1 seer; 40 s. = 1 maund; 3 m. = 2 cwt., - - -	= 224.—
4 chattacs = 1 pouah; 5 seers = 1 pussaree.	
1 maund (*bazaar weight*) = 100 troy lbs., nominally, - - - - - -	= 82.133
11 factory maunds = 10 bazaar maunds.	
Madras (Carnatic, Coromandel coast): 10 pagodas *or* varahuns = 1 polam; 8 p. = 1 seer; 5 s. = 1 vis *or* visay; 8 v. = 1 maund *or* maon; 20 m. = 1 candy *or* baruay, - -	= 500.—
Goa: 32 seers = 1 maund; 20 m. = 1 bahar, -	= 495.—
Pondicherry: 10 varahuns = 1 poloin; 40 p. = 1 vis; 8 v. = 1 maund; 20 m. = 1 candy, -	= 588.—
Surat: 16 pice = 2 tipprees = 1 seer; 40 s. = 1 maund; 20 m. = 1 candy, - - -	= 300.—
3 candies = 1 bahur.	
Tatta: 4 pice = 1 anna; 16 a. = 1 seer = 72¼ tola; 40 s. = 1 maund, - - -	= 72.32
Mysore, Seringapatam, and Malabar coast generally: 40 polams = 1 vis *or* pussaree; 8 v. = 1 maund *or* maon, - - - -	= 30.—
20 maunds = 1 bahur *or* candy; 20 b. = 1 garce.	

Foreign.	*U. States.* Avoirdupois pounds.
Tranquebar, and Coromandel coast generally; 30 chittacks = 1 vis; 6⅗ v. = 1 maund; 20 m. = 1 candy, - - - - -	= 500.—
MALAYA (Malay Peninsula, Strait of Malacca): Same as at Singapore Island; also, 1 kip *for tin* = 40½ Av. lbs., and 20 buncals = 1 catty *for gold and silver* = 2¼ lbs., troy.	
PENANG ISLAND, *or* PRINCE OF WALES ISLAND (Areca Island, Strait of Malacca): Same as at Singapore.	
SIAM (Farther India): 2 tical = 1 tael; 20 t. = 1 catty; 100 c. = 1 pecul, - - -	= 135.2536
20 piculs = 1 cajar.	
SINGAPORE ISLAND (off S. extremity of Malay Peninsula): 16 tael = 1 catty; 100 c. = 1 pecul; 3 p. = 1 bahar, - - -	= 405.—
BANCA ISLAND (Malay Archipelago): Same as at Batavia, JAVA ISLAND.	
BORNEO ISLAND. — Same as at Batavia, JAVA.	
CELEBES ISLAND. — Same as at Batavia.	
JAVA ISLAND. — *Batavia, &c.*: 16 tael = 1 catty; 100 c. = 1 pecul; 3 p. = 1 bahur = 200 goelak, - - - - -	= 406.888
4½ peculs (300 goelaks) = 1 great bahar.	
MOLUCCAS *or* SPICE ISLANDS. — *Amboyna, the Banda Islands, Batshian, Booro, Ceram, Gilolo, Oby, Waigeoo*: Official, as at Batavia.	
PHILIPPINE ISLANDS (*Luzon, Mindano, Palawan, Mindoro, Panay, Marindique, Negros, Bohol, Zuba, Samar, Masbata, Leyte, &c.*): 100 catties = 1 pecul = 37½ Castilian libra, -	= 139.4957
1 caban of rice (*usual*), - - -	= 133.—
1 caban of cocoa, - - - -	= 83.5
SOOLOO ISLANDS (*Sooloo, Basseelan, Tawee-Tawee, Pilas, Pala, Tapul Isles, &c.*): 10 mace = 1 tael; 16 t. = 1 catty; 50 c. = 1 lachsa; 2 l. = 1 pecul, - - - - -	= 133⅓.—
SUMATRA ISLAND. — 16 mace = 1 tael; 25 t. = 1 catty; 36 c. = 1 maund; 5½ m. = 1 candil *or* bahur, - - - - -	= 423.5
4 catties = 5 goelaks. 1 tael = 133⅓ lbs., Av.	
14⅔ salup = 7⅓ ootan = 1 nelli, - -	= 29⅓.—

Foreign.	*U. States.* Avoirdupois pounds.
ITALY. — The metric system of weights, either under the French denominations or as follows, is now the official, and may be considered the general commercial system throughout Italy, the islands of Sardinia and Sicily included.	
10,000 grani = 1000 denari = 100 grossi = 10 oncie = 1 libbra = 1 kilogram.	
1000 libbre = 100 rubbi = 10 quintali *or* centinaji = 1 migliajo *or* 1 tonnellata, - -	= 2204.6214
Special and local:	
Carrara: 1 cubic palmo of marble = 884.74 cubic inches, - - - - - -	= 90.82
JAPAN (N. Pacific Ocean). — *Niphon I., Kioo-Sioo I., Sikokf I., the dependencies Yeso I., Bonin I., the Loo-Choo group, &c.:*	
1000 moo = 100 rin = 10 sen = 1 monme = 26.784 troy grains.	
160 monme = 1 kan, - - - -	= 0.612206
350 monme ($2\frac{3}{16}$ kan) = 1 catty; 100 c. = 1 pecul, - - - - -	= 133.92

NOTE. — In commercial transactions the pecul is usually reckoned at 133⅓ lbs., the same as in China; but it is equal to 133.92 lbs. by the weights and analyses of the modern Japanese coins.

LIBERIA (W. Africa): Same as in the United States.	
MADEIRA ISLANDS (Atlantic Ocean, off W. coast of Morocco, N. Africa): Same as in Portugal.	
MALTA ISLAND (Mediterranean Sea, S. of Sicily): Official, same as in Great Britain.	
MAURITIUS ISLAND (Mascarene group, Indian Ocean): Official, as in Great Britain; also, 100 livres (old poids de marc of France) = 1 quintal, - - - - -	= 107.9184
MEXICO. — 8 ochavas = 1 onza; 16 o. = 1 libra; 25 l. = 1 arroba; 4 a. = 1 quintal, -	= 101.4232
MOROCCO (Barbary, N. Africa): 10 onza = 1 mark; 2 m. = 1 rotl; 100 r. = 1 cantaro -	= 118.65664

NOTE. — The commercial rotl of Morocco, both in theory and practice, is equal to the weight of 20 old standard duros (silver dollars) of Spain; the miner's rotl is equal to 100 meticals (1 old Venetian libbra, peso grosso), or 1.054945 avoirdupois pound; and the market rotl = 160 meticals.

Foreign.	*U. States.* Avoirdupois pounds.
MOZAMBIQUE (E. Africa): Same as in Portugal.	
NORWAY. — Same as in Denmark.	
NUBIA (E. Africa): Same as at Alexandria, Egypt.	
PERSIA. — 6 dirhem = 2 mascais = 1 miscal = 73.846154 troy grains.	
100 miscals = 1 rotl *or* ratel, - - - =	1.054945
136½ rotl = 144 avoirdupois pounds.	
6½ rotl = 1 maund tabree (customs), - - =	6.85714
6 rotl = 1 " " (bazaar), - - =	6.3297
7½ rotl = 1 maund copra (customs), - - =	7.91209
7 rotl = 1 " " (bazaar), - - =	7.3846
Also, 116⅔ miscals = 1 rotl copra (bazaar), and 6 rotl copra = 1 maund copra, bazaar.	
2 maunds tabree (of Tabreez) = 1 maund shah (of Sheeraz).	
1600 miscals = 1 reh of Teheran = 16 bazaar rotl of Tabreez = 6 okes of Turkey, - =	16.88023

NOTE. — The maund tabree is used chiefly for weighing coarse metals, coffee, sugar, drugs, &c., and the maund copra for weighing rice and provisions. The maund shah is used chiefly in Sheeraz, Bushire, and Gombroon, although the last-mentioned port now belongs to the Muscat dominion.

PORTUGAL. — 576 grao = 24 escropulo = 8 outava = 1 onça; 16 o. = 2 marco = 1 arratel, =	1.01187
32 arratel = 1 arroba; 4 arroba = 1 quintal, - =	129.5193
13½ quintalo = 1 tonelada.	
ROME and *Civita Vecchia:* 24 grao = 1 oncia; 12 o. = 1 libbra; 10 l. = 1 decime; 10 d. = 1 centinajo *or* cantaro, - - - =	74.7714
10 centinajo = 1 migliajo.	
RUSSIA. — 96 solotnik = 32 loth = 12 lana = 1 funt, - - - - - - =	0.902612
40 funt = 20 dowinik = 13⅓ trowinik = 8 paterik = 4 desaterik = 1 pud, - - =	36.10448
10 puds = 1 berkowitz; 3 b. = 1 paken; 2 p. = 1 last.	
SENEGAMBIA (W. Africa). — *Bathurst, Sierre Leone:* Official, as in Great Britain.	
St. Louis: Official, as in France.	
SOCOTRA ISLAND (Indian Ocean, off E. coast of Africa): Same as at Muscat, Arabia.	
SPAIN. — *Standard of Castile:* 128 ochavas = 16 onzas = 2 marcos = 1 libra, - - =	1.0145143

Foreign.	*U. States.* Avoirdupois pounds.
25 libras = 1 arroba; 4 a. = 1 quintal; 20 q. = 1 tonelada, - - - -	= 2029.028
Special and local, but not official:	
VALENCIA. — *Alicante, &c.:* 12 onze = 1 libra menor (*minor*).	
18 onze = 1 libra mayor (*major*); 24 l. mayor, *or* 36 l. menor, = 1 arroba = $1\frac{1}{8}$ Castilian arroba, - - - - - - -	= 27.3919
4 arrobas = 1 quintal; $2\frac{1}{2}$ q. = 1 carga; 8 c. = 1 tonelada, - - - -	= 2191.291
ASTURIAS. — *Santander, &c.:* 25 libras = 1 arroba = $1\frac{1}{2}$ arroba of Castile, - - -	= 38.04429
ARAGON. — *Saragossa, &c.:* 36 libras = 1 arroba,	= 27.3919
BISCAY. — *Bilboa, &c.:* 25 libras = 1 arroba = $1\frac{1}{16}$ Castilian arroba, - - - -	= 26.948
146 libras = 1 quintal macho (*for iron*).	
CATALONIA. — *Barcelona, &c.:* 25 libras = 1 arroba = $\frac{7}{8}$ Castilian arroba, - - -	= 22.1925
ANDALUSIA. — *Malaga:* 7 arrobas (*Castilian*) = $1\frac{3}{4}$ quintal = 1 carga of raisins, - -	= 177.54

NOTE. — The employment of the metric system of weights is sanctioned by law in Spain.

SWEDEN. — *New standard:* 100 korn = 1 ort; 100 o. = 1 skalpund; 100 s. = 1 centner, -	= 92.8583
SWITZERLAND. — *New system:* 32 lothe, *or* 8 gros, = 1 unze; 16 u. = 1 pfund *or* livre = $\frac{1}{2}$ kilogram; 100 pfunds = 1 centner, -	= 110.2311
TRIPOLI (Barbary, N. Africa.) — 16 karob = 1 drachma; 10 d. = 1 oncia *or* usano = $6\frac{1}{2}$ miscals *or* 1 troy ounce.	
16 oncia = 1 rotl; 100 r. = 1 cantaro, -	= 109.7143
400 drachma = 1 oka, - - - -	= 2.742857
TUNIS (Barbary, N. Africa). — Same as in Tripoli.	
TURKEY. — 4 grani = 1 kara, killot, *or* taim; 16 k. = 1 dirhem *or* drachmia = $\frac{2}{3}$ miscal *or* metical = $49\frac{3}{13}$ troy grains.	
100 dirhems = 1 cheki; 4 c. = 1 oka, -	= 2.813187
250 dirhems = 1 cheki for opium = $166\frac{2}{3}$ miscals, - - - - - -	= 1.75836
Constantinople, Galata: 2 cheki = 1 rotl *or* rotolo; 2 r. = 1 oka; 50 o. = 1 kantar *or* cantaro grosso, - - - - -	= 140.65934

Foreign.	*U. States.* Avoirdupois pounds.
176 dirhems = 1 rotl; 100 r. = 44 okes = 1 cantaro sottile, - - - - =	123.78834
610 dirhems = 1 teffeh of Brusa silk, - - =	4.29039
800 dirhems = 1 teffeh of goat's wool.	
Smyrna: 180 dirhems = 120 miscals = 1 rotolo; 100 r. = 45 okes = 1 cantaro, - - =	126.60171
44 okes = 1 cantaro for tin.	
In Bassorah the Arabs commonly employ the light wakea of 160 krat of Mocha = 1 troy ounce, or 480 grains; and 15 wakea = 1 cheki, - =	1.0285143
40 wakea ($2\frac{2}{3}$ cheki) = 1 oka, - - - =	2.742857
50 okes (2,000 wakea) = 1 cuttra *or* cantaro, =	137.142857

NOTE.—The initial for the avoirdupois values of the Turkish weights, in the absence of documentary statistics on the subject, if any exist, was derived from the Abyssinian dirhem and by comparison; and the result, I find, is almost strictly confirmed by assays carefully made at the United-States mint and elsewhere, of the modern gold and silver coins of Ottoman mintage compared with their present official standards; viz., $\frac{1}{320}$ cheki of fine silver to the piastre and $\frac{1}{4850}$ cheki of fine gold to the piastre. The *troy ounce*, it is well known, was derived from the Abyssinian dirhem (drachma) or its multiple by 10, the wakea, vakia, or wakega, and consists of 12 of the first-mentioned units, making the dirhem equivalent to 40 troy or United-States grains, while 120 of these dirhems, or 1 rotl or rotolo of Abyssinia, is equal to 65 miscals, or meticals, or $\frac{1}{10}$ maund tabreė (customs) of Persia; hence, $120 \times 40 \div 65 = 73\frac{11}{13}$ troy grains, the value of the Persian miscal. But the miscal, or metical of Persia, and that of Turkey are alike: in theory it is the same specific weight everywhere; and 1 dirhem of Turkey is equal to $\frac{2}{3}$ miscal; hence, $73\frac{11}{13} \times \frac{2}{3} = 49\frac{3}{13}$ troy grains, the value of the Turkish dirhem, and 4 dirhem of Persia are equal to 1 dirhem of Turkey, and $6\frac{1}{2}$ miscals are equal to 1 troy ounee.

McCulloch (not to go farther back), in his work published in 1839, says the cantaro of Constantinople of 45 okas is equal to 127.2 avoirdupois pounds; or, in other words, that the oka of Constantinople is equal to 2.82667 pounds; and he states the oka of Smyrna to be equal to 2 lbs. 13 oz. 5 dr., or 2.83203 pounds, but, at the same time, under the last-mentioned head (Smyrna), states the weights and measures to be the same as those of Constantinople.

Alexander, in his work published in 1850, places the oka of Constantinople at 2.828571 pounds, and that of Smyrna at 2.812488 pounds, in a measure reversing the values by McCulloch; while Noback, in a work of more recent date, says the oka of Constantinople is equal to 1278.48 grammes = 2.8185644 pounds; making it nearly equal to that of Smyrna by Alexander; and that the oka of Smyrna is a little heavier, being equal to 2.83286 pounds.

From these conflicting statements no tenable idea can be gained except this; viz., that the initial and leading weights of Asia Minor (Anatolia) are probably theoretically and practically the same as those of Turkey in Europe. But this seems to admit of no question, since 1 batman of Persian silk, containing 1 reh, or 1600 miscals of Teheran, is invariably equal, both in Constantinople and Smyrna, to 6 okas of Turkey; wherefore, the oka is equal to $\frac{1600}{6} = 266\frac{2}{3}$ miscals, or $\frac{266\frac{2}{3} \times 73\frac{11}{13}}{7000} = 2\frac{74}{91}$ avoirdupois pounds, being slightly heavier than that of Smyrna by Alexander, and a trifle lighter than that of Constantinople by Noback.

Foreign. *U. States.*

West Indies.

Avoirdupois pounds.

GREAT ANTILLE ISLANDS. — CUBA: Standard of Castile. 16 onzas = 1 libra; 25 l. = 1 arroba; 4 a. = 1 quintal; 20 q. = 1 tonelada, - . - - - = 2029.028

HAYTI: Poids du marc of France, previous to A.D. 1800.

16 onces = 1 livre; 100 l. = 1 quintaux; 10 q. = 1 millier, - - - - - = 1079.176

2 milliers *or* barriques = 1 tonneau.

SAN DOMINGO, OR DOMINICA: Same as in Cuba.

PORTO RICO: Same as in Cuba.

JAMAICA: 16 ounces = 1 pound; 28 p. = 1 quarter; 4 q. = 1 cwt.; 20 cwt. = 1 ton, - = 2240.—

LUCAYOS, OR BAHAMA ISLANDS. — Same as in Jamaica.

CARIBBEE ISLANDS. *LEEWARD GROUP.* — DOMINICA, TORTOLA, VIRGIN GORDA, ST. CHRISTOPHER, ANGUILLA, BARBUDA, NEVIS, SABA: Same as in Jamaica.

ANTIGUA, MONTSERRAT: 100 pounds = 1 cental *or* cwt., - - - - - = 100.—

ST. EUSTACIUS: Official, as in Holland.

GUADELOUPE, MARIE GALANTE, DESIRADE, LES SAINTES: Official, as in France; also as in Hayti.

ST. MARTIN: Dutch, as in Holland; French, as in France; also as in Hayti.

ST. THOMAS, SANTA CRUZ, ST. JAN: Official, as in Denmark.

ST. BARTHOLOMEW: Official, as in Sweden.

WINDWARD GROUP. — BARBADOES, GRENADA, ST. VINCENT, TOBAGO: Same as in Jamaica.

MARTINIQUE, ST. PIERRE, ST. LUCIA: Same as in Hayti.

TRINIDAD: Same as in Cuba; official, as in Great Britain.

LITTLE ANTILLES. — CURACOA, BUEN AYRE, ORUBA: Official, as in Holland.

MARGARITA, TORTUGA, BLANQUILLA: Same as in Venezuela.

ZANGUEBAR (E. AFRICA). — *Zanzibar* (Island and Town). Consul's report: 12 maunds = 1 frasler, - - - - - = 35.—

FOREIGN LIQUID MEASURES,

REDUCED TO THEIR EQUIVALENT VALUES IN THE UNITED STATES.

Foreign.	*U. States.* Wine gallons.
ABYSSINIA. — By weight: see Weights.	
ALGERIA. — Official, as in France; also, 16⅔ litres = 1 khoulle; 6 k. = 1 hectolitre, - - =	26.417
ARABIA. — (Generally by weight.) *Mocha:* 20 wakeas (weight) = 1 nusfiah; 8 n. = 1 cuda *or* gudda = 16 av. lbs.	
AUSTRALASIA. — Same as in Great Britain.	
AUSTRIA (*legal for the Empire*): 4 seidel = 1 maass; 10 m. = 1 viertel; 4 v. = 1 eimer *or* orna, - - - - - - =	14.9543
32 eimers = 1 fuder; 42½ maass = 1 eimer for beer.	
AZORE ISLANDS. — Same as in Portugal.	
BALEARIC ISLANDS. — MAJORCA: 8 quartas = 1 quartera; 3½ quarteras, *or* 4 quartinellos, = 1 quartin *or* barril; 4 quartin = 1 carga; 4 c. = 1 botta = 27 fluid arrobas, or 1 pipa of Castile, - - - - =	114.9692
MINORCA: 8 quartas = 1 quartera; 4 quarteras, *or* 2⅔ gerah, = 1 quartin *or* barril; 4 barrils = 1 carga; 4 c. = 1 botta = 31⅓ fluid arrobas of Castile, - - - - =	133.1635
BELGIUM. — Same as in France.	
BERBERA. — Same as at Mocha, Arabia.	
BERMUDA ISLANDS. — Official, as in Great Britain; in trade, generally as in the United States.	
BOURBON ISLAND. — Same as in France.	
CANADA, DOMINION OF. — Official, as in Great Britain; in trade, as in the United States.	
CANARY ISLANDS. — Same as in Spain, Castilian Standard.	
CANDIA ISLAND. — 1 mistata = 8½ okes weight, or, of olive oil, - - - - =	2.9397
CAPE OF GOOD HOPE. — Same as in Great Britain.	
CAPE VERDE ISLANDS. — Same as in Portugal.	

Foreign.	*U. States.*
Central and South America.	Wine gallons.
GUATEMALA, HONDURAS, SAN SALVADOR, NICARAGUA, COSTA RICA: Same as in Spain, standard of Castile; also the wine gallon of the United States is used.	
BALIZE: Official, as in Great Britain.	
BRAZIL: 24 quartilhos = 12 garrafa = 6 canada = 3 medida = 1 alqueire *or* pote = 18 arratels weight, - - - - - - =	2.18418
60 potes = 1 pipa = 1080 arratels weight, - =	131.051
Bahia: 1 canada = 15½ arratels weight = 5⅛ canadas of Rio Janeiro, - - - =	1.88082
72 canadas = 1 pipa of spirits, - - - =	135.4193
100 canadas = 1 pipa of molasses.	
ARGENTINE REPUBLIC: 4 cuartos = 1 frasco; 8 f. = 1 caneca = 19 litres in theory, - - =	5.01927
3 frascos = 1 cortan; 16 c. (6 canecas) = 1 carga, - - - - - =	30.11541
4 cargas = 1 pipa catalana; also, 8 frascos = 5 U. S. gallons, nominally.	
PERU, CHILI, BOLIVIA, ECUADOR, NEW GRANADA, VENEZUELA, URUGUAY, PARAGUAY: Chiefly as in Spain, standard of Castile.	

NOTE.—In the States last mentioned, the U. S. wine gallon is more or less used in trade; and in Chili it is the customs' unit-measure for liquids. Also, in Chili, Peru, New Granada, Bolivia, and Venezuela, the use of the metric system is sanctioned by law, and may be expected to gradually come into use.

GUIANA — *Cayenne:* Same as in France.	
Paramaribo (Surinam): Same as in Holland; also as in Cayenne.	
Georgetown (Demerara): Same as in Great Britain.	
FALKLAND ISLANDS: Same as in Great Britain.	
CHINA. — By weight only; for denominations, see DRY MEASURES.	
DENMARK. — 42 pagel = 4 potte = 2 kande = 1 stubchen, - - - - - - =	1.02089
40 stubchen = 20 viertel = 4 anker = 1 ohm, =	40.83522
1½ ohm = 1 oxehoved; 2 oxehoved = 1 piba; 2 p. = 1 fuder; 1¼ f. = 1 stykfad.	
34 stubchen = 1 toende *for beer;* 30 stubchen = 1 toende *for tar.*	
EGYPT. — By weight exclusively. See WEIGHTS.	

Foreign. *U. States.* Wine gallons.

FRANCE.—1,000 millilitres = 100 centilitres =
10 decilitres = 1 litre, - - - = 0.26417
100 litres = 10 decalitres = 1 hectolitre, - = 26.417029
100 hectolitres = 10 kilolitres = 1 myrialitre.

GERMANY.—Prussian and Zollverein maasse for
all the States of the tariff-alliance: 2 oessel
= 1 quart; 30 q. = 1 anker; 2 a. = 1 eimer
= 3840 cubic Rhine zollen; or, since $38\frac{1}{4}$ zollen = 1 metre, = $68\frac{2212764}{3581577}$ litres, - - = 18.126789
2 eimers = 1 aam, ahm, *or* ohm; 3 eimers =
1 oxhoft; 12 eimers = 1 fuder.
$3\frac{1}{3}$ eimers = 2 barrile = 1 fass *for beer*, - = 60.42263

NOTE.—I have been thus particular in treating of the eimer, because the notion seems to be generally entertained that it is equal in theory to 68.7 litres.

Special and local, or domestic:

BADEN: 1000 maass = 100 stubchen = 10 ohm =
1 fuder = 15 hectolitres, - - - = 396.2555

BAVARIA: 4 quartile = 1 mäss, *or* mässkanne; 60
m. = 1 eimer, - . - - - = 16.9452
25 eimer = 1 fass; 64 mässe = 1 eimer *for beer.*

BREMEN: 4 mengel = 1 quartier, *or* vierling; 4 q.
= 1 stubchen, - - - - = 0.85106
$2\frac{1}{4}$ stubchen = 1 viertil; 5 v. = 1 anker; 4 a.
= 1 ahm, - - - - - = 38.29781
6 ankers = 1 oxhoft; 4 o. = 1 fuder; 44 stubchen = 1 ahm *for wine.*
6 stubchen = 1 steckannen; 6 s. = 1 tonne *for
train oil* = 216 pfunde weight, or, at $7\frac{3}{4}$ av.
lbs to the gallon, - - - - = 30.6073

BRUNSWICK: 10 stubchen = 1 anker; 4 a. = 1
ahm; $1\frac{1}{2}$ ahm = 1 oxhoft, - - - = 59.2803

HAMBURG: 8 oessel, plank, *or* stück = 4 quartier,
or potts = 2 kannen = 1 stubchen, - - = 0.956404
8 stubchen = 4 viertel = 1 eimer, - - = 7.651224
32 stubchen = 1 anker; 40 stubchen = 1 ohm;
60 stubchen = 1 oxhoft; 240 stubchen = 1
fuder.

HESSE DARMSTADT: 16 schoppen = 4 masschen
= 1 viertel; 20 v. = 1 ohm = 16 decalitres, = 42.26725

LUBEC: 16 ort = 8 plank, *or* nossel = 4 quartier
= 2 kanne = 1 stubchen, - - - = 0.9546
8 stubchen = 4 viertel = 1 eimer, - - = 7.6512
5 viertels = 1 anker; 6 a. = 1 fass, - - = 57.384

Foreign.	*U. States.* Wine gallons.
MECKLENBURG (*legal*): Same as in Hamburg.	
OLDENBURG: 240 quartiers, *or* 156 kannies = 1 oxhoft (*legal*) = 1 fuder of Lubec.	
SAXONY (*legal*): 144 nössel = 72 kanne = 24 viertel = 1 eimer, - - - -	= 17.8107
2 eimers = 1 aam; 3 eimers = 1 oxhoft; 5 eimers = 1 fass; 12 eimers = 1 fuder.	
WURTEMBERG. — *Helleich mäss:* 4 quartier, *or* schoppen = 1 maas; 10 m. = 1 immer; 16 i. = 1 eimer; 6 e. = 1 fuder, - - -	= 465.9036
GREAT BRITAIN. — *Imperial measure:* Denominations and relative values same as in the United States, but capacity values = $20\frac{74}{231}$ per cent greater. See LIQUID MEASURES, U. S.	
1 imperial gallon = $\frac{277274}{231000}$ or 1.200320344 wine gallons of the United States.	
GREECE. — 1 kila, *or* galloni = $2\frac{1}{2}$ okas weight.	
HOLLAND. — 10 vingerhoed = 1 maatje; 10 m. = 1 kan; 100 k. = 1 vat = 1 hectolitre, -	= 26.417
HAWAIIAN ISLANDS. — Same as in the United States.	

India and Malaysia, or East Indies.

ANNAM, BURMAH, *Calcutta*, and BENGAL generally, CEYLON I., PHILIPPINE IS., SOO-LOO IS. — By weight. See WEIGHTS.	
Bombay, Madras: By weight, chiefly; the wine gallon of the United States is sometimes used.	
Goa: Same as in Portugal.	
Pondicherry: Official, as in France.	
MALACCA. — Capacity measures, same as in the United States.	
PENANG ISLAND. — Same as in Singapore.	
SIAM. — 20 canan = 1 cohi = 80 catties weight, or 108.203 av. lbs., - - - -	= 12.9757
SINGAPORE ISLAND. — Capacity measures, same as those of the United States.	
BANCA I., BORNEO I., CELEBES I., JAVA I., MOLUCCA IS., SUMATRA I. — Official, same as in Holland.	
ITALY. — The metric measures of capacity are used here, both under the French nomencla-	

Foreign.	*U. States.* Wine gallons.
ture and as follows; viz., 10 coppa = 1 pinta; 10 p. = 1 mina; 10 m. = 1 soma = 1 hectolitre, - - . - - =	26.417
JAPAN. — 1 tsjoo = $\frac{1}{16}$ cubic kani-sasi = 106.09663 cubic inches, - - - - - =	0.459293
10 tsjoo = 1 to; 10 to = 1 kok; 10 sasi = 1 goo; 10 goo = 1 tsjoo.	
LIBERIA. — Same as in the United States.	
MADEIRA ISLANDS. — Same as in Portugal.	
MALTA ISLAND. — Official, same as in Great Britain; also, 1 caffiso of oil = 4.724 gallons, and 1 barrile of wine = 9.448 gallons.	
MAURITIUS ISLAND. — 8 pintes = 1 velt, - =	1.969
MEXICO. — Chiefly as in Spain, *Castilian standard;* but the use of the metric system is legalized, and may be expected soon to be introduced into practice.	
MOROCCO. —	
MOZAMBIQUE. — Same as in Portugal.	
NORWAY. — The Danish capacity measures are used here.	
NUBIA. — By weight, as at Alexandria.	
PERSIA. — By weight. See WEIGHTS.	
PORTUGAL. — 24 quartilhos = 6 canadas = 1 alqueire, *or* pote = 18 arratels weight, - =	2.18418
2 potes = 1 almude; 26 a. = 1 bota *or* pipa, - =	113.5775
2 botas = 1 tonelada; 18 almudes = 1 barril.	
Oporto: 1 alqueire = 27¼ arratels weight, or $\frac{100}{66}$ alqueires of Lisbon, - - - - =	3.30936
ROME, and *Civita Vecchia:* 64 cartocci = 16 quartucci = 4 foglietti = 1 boccale, - =	0.48165
32 boccali = 1 barile; 16 b. = 1 botta, - =	246.605
32 boccale *for wine* = 28 boccale *for oil.*	
RUSSIA. — 100 tscharka = 10 krushka = 1 vedro, *or* wedro, - - - - - =	3.24674
3 vedros = 1 anker; 6 a. = 1 oxhoft, - =	58.4413
40 vedros = 1 botschka.	
SPAIN (*Castilian standard*): 4 copas = 1 cuartillo; 4 c. = 1 azumbra; 8 a. = 1 arroba *or* cantaro = 35 libras weight of distilled water at maximum density, or 35.508 av. lbs., =	4.25812
16 arrobas = 1 moyo; 27 arrobas = 1 pipa; 30 arrobas = 1 bota; 60 arrobas = 1 tonelada.	
1 arroba menor *for oil* = 27¼ libras weight, - =	3.31525

Foreign.	*U. States.* Wine gallons.
Special and local:	
Alicante and Valencia: 16 cuartillos = 4 cuartos = 1 arroba = $\frac{3}{4}$ Castilian arrobas, - -	= 3.19359
40 arrobas = 1 pipa; 2 p. = 1 tonelada, -	= 255.4872
Also 100 cantaros of $\frac{8}{5}$ Castilian arroba each = 1 tonelada.	
Barcelona: 16 cortans (12 arrobas Catalan weight) = 1 carga = 7½ fluid arrobas of Castile, - - - - -	= 31.9359
Gibraltar: 38 arrobas menor of Castile = 1 pipa,	= 125.9795
126 U. S. gallons, or 105 imp. gallons in theory = 1 pipa.	
Malaga: 33⅓ fluid arrobas of Castile = 1 pipa, -	= 141.9373

NOTE. — The employment of the metric capacity measures is sanctioned by law in Spain.

SWEDEN. — 4 qwarter = 2 stop = 1 kanna = $\frac{1}{10}$ cubit fot.	
48 kannas = 8 ottingar = 4 fjerding = 1 tunna, - - - - - - -	= 33.184106
SWITZERLAND. — (*Official and legal for the 22 Cantons*): 1000 emine = 100 maass *or* potts = 10 gelt = 1 saum = 150 litres, - -	= 39.62555
TRIPOLI. — 14 caraffa = 1 mataro *for oil* = 17½ okas weight, or 48 av. lbs.	
40 caraffa = 1 barril = 50 okas weight; also 24 bozza = 1 barile = 130 rotolos weight (52 okas), or, of the standard of the United States, - - - - -	= 17.10402
TUNIS. — 2 mettars for wine = 1 mettar for oil = 36 rotoli weight; 3⅓ mettars = 1 millerolle = 120 rotoli weight for oil, or 231.65716 av. lbs.	
TURKEY. — By weight. See WEIGHTS. Also 1 almud of oil = 8 okas.	

West Indies.

Cuba, Porto Rico: Same as in Spain, Castilian standard; but in Cuba the U. S. gallon is also used: 36 gallons = 1 bocoy = 36 U. S. gallons.

Dominica (*San Domingo, or Dominican Republic,* HAYTI I.): Same as in Spain, standard of Castile.

Foreign.	*U. States.* Wine gallons.
Hayti, Empire of (HAYTI I.). — 60 gallons = 1 tierçon, - - - - - -	= 60.—

Guadeloupe, Martinique, Marie Galante, Les Saints, Desirade, northern portion of *St. Martin.* — Official, as in France; but in trade the United-States fluid gallon is chiefly used; *for molasses,* 30 gallons = 1 barāl; 65 gallons = 1 tierçon; 105 gallons = 1 baucaut: *for rum,* 114 gallons = 1 boucaut.

Jamaica, Trinidad, Bahamas, Barbadoes, St. Christopher, Dominica, Montserrat, Grenada, St. Lucia, Antigua, Tortola, Tobago, Nevis, Virgin Gorda, Grenadines: Official, as in Great Britain; in trade, mostly as in the United States.

St. Thomas, Santa Cruz, St. Jan: Official, as in Denmark.

St. Eustatius, Curacoa, Buen Ayre, Oruba, southern portion of *St. Martin:* Official, as in Holland.

Margarita, Tortuga, Blanquilla: Same as in Venezuela.

St. Bartholomew: Official, as in Sweden.

FOREIGN DRY MEASURES,

REDUCED TO THEIR EQUIVALENT VALUES IN THE UNITED STATES.

Foreign.	*U. States.* Winchester bushels.
ABYSSINIA. — 24 madega = 1 ardeb, - - =	0.33333
ALGERIA. — Official, as in France; also 16 tarrie = 8 saa *or* saha = 1 caffiso, - - =	9.—
ARABIA. — By weight: 40 kellas = 1 tomaun *for rice* = 56 maunds weight, or 168 av. lbs.	
AUSTRALIA. — Same as in Great Britain.	
AUSTRIA (*legal for the Empire*). — 4 becher = 1 mässel; 4 m. = 1 viertel; 4 v. = 1 metze, - =	1.7452
AZORE ISLANDS. — 16 quartos = 4 alqueires = 1 fanga; 15 f. = 1 moio = $\frac{7}{8}$ moio of Lisbon, - - - - - - =	20.5298
BALEARIC ISLANDS. — 6 barcella = 1 quartera = $1\frac{1}{3}$ fanega of Castile, - - - =	2.157
BELGIUM. — 100 kop = 10 schepel = 1 mudde = 1 hectolitre, - - - - =	2.83774
BERBERA. —	
BERMUDA ISLANDS. — Official, as in Great Britain; the U. S. bushel is also used.	
BOURBON ISLAND. — Same as in France.	
CANADA, DOMINION OF. — Official, as in Great Britain; in trade, as in the United States, except that in Lower Canada the old French minot = 1.107436 U. S. bushels is used.	
CANARY ISLANDS. — 12 celamins = 1 fanega = 136 libras weight, - - - =	1.77737
$16\frac{2}{3}$ celamins = 1 fanega heaped.	
CANDIA ISLAND. — 1 carga, - - - =	4.3211
CAPE OF GOOD HOPE. — Same as in Great Britain; also, 4 schepels = 1 muid - - =	3.1564
CAPE VERDE ISLANDS. — Same as in Portugal.	

Central and South America.

GUATEMALA, HONDURAS, SAN SALVADOR, NICARAGUA, COSTA RICA: Same as in Spain, Castilian standard; but the U. S. bushel is also used.

Foreign.	*U. States.* Winchester bushels.
BALIZE. — Official, as in Great Britain.	
BRAZIL. — 16 quartas = 4 alqueirs = 1 fanga; 15 f. = 1 moio = $\frac{6}{7}$ moio of Portugal, - -	= 20.11086
Bahia: 1 alqueire = 67½ arratels weight, or 2¼ alqueires of Portugal, - - - -	= 0.87985
Maranham: 1 alqueire = 100 arratels weight, -	= 1.30348
ARGENTINE REPUBLIC, URUGUAY. — 4 cuartillos = 1 fanega = 134 litres, - - -	= 3.80257
CHILI. — 12 celamins = 1 fanega, - - -	= 2.5753
PERU. — 1 fanega, - - - - -	= 2.31777
BOLIVIA, ECUADOR, NEW GRANADA, VENEZUELA, PARAGUAY: Chiefly as in Spain, Castilian standard.	
GUIANA. — *Cayenne:* Same as in France.	
Paramaribo: Same as in Holland, also as in Cayenne.	

NOTE. — The use of the metric system is sanctioned by law in Chili, Peru, New Granada, Bolivia, Venezuela, and French and Dutch Guiana, and, to some extent, is introduced into practice.

Georgetown: Same as in Great Britain.	
FALKLAND ISLANDS. — Same as in Great Britain.	
CHINA. — 10 hŏ = 1 shing; 10 s. = 1 tau; 10 t. = 1 hwŭh, seí, *or* tane = 120 catties weight = 160 av. lbs.	
DENMARK. — 32 sextingkar = 16 ottingkar = 2 skieppe = 1 fjerding, stubchen, *or* scheffel = 36 potte, - - - - -	= 0.98698
4 fjerding = 1 toende; 22 t. = 1 last, - -	= 86.8546
EGYPT. — *Cairo:* 24 robi = 6 usbek = 1 ardeb = 144 okas weight, - - - -	= 5.088
Alexandria, Rosetta: 1 kislos = 137 okas weight,	= 4.84065
1 rebeb = 126 okas weight, - - -	= 4.45198
1 ardeb = 230 okas weight, - - -	= 8.12663
FRANCE. — 100 litres = 10 decalitres = 1 hectolitre, - - - - - -	= 2.83774
100 hectolitres = 10 kilolitres = 1 myrialitre, -	= 283.774
GERMANY. — PRUSSIA, and Zollverein mäss of all the States of the tariff-alliance: 16 metzen (3072 cubic Rhein zollen, or 48 fluid zollverein quarts) = 1 scheffel, - - -	= 1.55776
Special and local:	
BADEN: 1000 becher = 100 mässlein = 10 sester = 1 malter = 1½ hectolitre, - - -	= 4.25661
10 malters = 1 zober.	

Foreign.	*U. States.* Winchester bushels.
BAVARIA: 16 dreissiger = 4 mässel = 1 viertel, -	= 0.525855
12 viertels ($17\frac{1}{3}$ fluid mässkanne, or six old metzen) = 1 scheffel, - - - -	= 6.310263
BREMEN: 16 spint = 4 viertel = 1 scheffel, -	= 2.10289
10 scheffels = 1 quarter; 4 q. = 1 last, -	= 84.11572
BRUNSWICK: 4 metzen = 1 himt; 40 h. = 1 wispel, - - - - - -	= 35.3514
HAMBURG: 8 spint = 2 himten = 1 fass = 1 zollverein scheffel, - - - -	= 1.55776
10 scheffels = 1 wispel; 6 w. = 1 last, -	= 93.4656
HESSE DARMSTADT: 64 kopfchen = 32 maasschen = 8 gescheid = 2 kumpf = 1 metze, -	= 0.45104
8 metzen = 2 simmer = 1 malter = 128 litres,	= 3.632308
LUBEC: 16 fass = 4 scheffels = 1 tonne, - -	= 3.9381
3 tonnen = 1 dromt; 8 d. = 1 last, - -	= 94.5139
MECKLENBURG: 16 spint, *or* metzen = 4 fass, *or* viertel = 1 scheffel, - - - -	= 1.103628
4 scheffels = 1 wispel; 3 w. = 1 last, - -	= 105.9483
OLDENBURG: 16 kannen = 1 scheffel; 8 s. = 1 tonne, - - - . . -	= 5.17536
$1\frac{1}{2}$ tonne = 1 molt; 12 m. = 1 last, - -	= 93.1565
SAXONY: 16 mässchen = 4 metzen = 1 viertel, -	= 0.73713
4 viertels = 1 scheffel; 12 s. = 1 malter, -	= 35.3823
2 malters = 1 wispel; 6 w. = 1 last, - -	= 424.5876
WURTEMBERG: 32 viertelein = 8 ecklein = 1 vierling, - - - . - -	= 0.157172
32 vierling = 8 simri = 1 scheffel, - -	= 5.0295
GREAT BRITAIN. — *Imperial measure:* Denominations and relative values same as in the United States, but capacity values = $\frac{554548}{537605}$ greater; 1 bushel = 1.0315157 U. S. bushels.	
GREECE. — 1 kila, - - - -	= 0.944
HOLLAND. — 1000 maatje = 100 kopen = 10 schepel = 1 mudde *or* zac = 1 hectolitre, -	= 2.83774
30 mudden = 1 last, - - - -	= 85.1322

India and Malaysia, or East Indies.

ANNAM, BURMAH, CEYLON ISLAND: By weight. See WEIGHTS.

HINDOSTAN. — *Bombay:* 8 tipprees = 4 seers = 1 adoulie.

Foreign.	*U. States.* Winchester bushels.
16 adoulies = 1 para = 8$\frac{3}{4}$ maunds weight, or 245 av. lbs., - - - - -	= 3.15607
8 paras = 1 candy = 70 maunds weight, -	= 25.248545
Calcutta: 80 chattac = 16 koonke = 4 raik = 1 pallie = 5 seers weight, or 9$\frac{1}{3}$ av. lbs.: 12 pallies = 1 morah = 1$\frac{1}{2}$ factory maunds weight, or 112 av. lbs.	
20 pallies = 1 soallee = 2$\frac{1}{2}$ factory maunds, or 186$\frac{2}{3}$ av. lbs.	
16 soallees (40 maunds, or 2986$\frac{2}{3}$ av. lbs.) = 1 kahoon, - - - - -	= 38.47397
Madras: 64 ollock = 8 puddy = 1 marcal.	
5 marcals = 1 para = 5$\frac{2}{5}$ maunds weight, or 135 av. lbs.	
80 paras = 1 garce = 432 maunds, or 10800 av. lbs., - - - - - -	= 139.12464
Goa: Same as in Portugal.	
Pondicherry: Same as in France.	
MALACCA. — The Winchester bushel is used, also the coyang of Siam.	
PENANG ISLAND. — Generally as in the United States.	
SIAM. — 40 sat = 1 sesti; 40 s. = 1 cohi; 65 c. = 1 coyang = 52 peculs weight, or 7033.1872 av. lbs., - - - - -	= 90.6009
SINGAPORE ISLAND. — Generally as in the United States.	
BANCA ISLAND, BORNEO ISLAND, CELEBES ISLAND, JAVA ISLAND, MOLUCCA ISLANDS, SUMATRA ISLAND. — Official, as in Holland.	
ITALY. — See LIQUID MEASURES:	
1 soma (hectolitre), - - - -	= 2.83774
JAPAN. — See LIQUID MEASURES:	
1 kok (6$\frac{1}{4}$ cubic kani-sasi), - - -	= 4.93376
LIBERIA. — Same as in the United States.	
MADEIRA ISLANDS. — Same as in Portugal.	
MALTA ISLAND. — Official, as in Great Britain; also, 1 salma rasa ($\frac{4}{5}$ salma colma), - -	= 8.2202
MAURITIUS ISLAND. — Official, as in Great Britain.	
MEXICO. — Chiefly as in Spain, standard of Castile; but the use of the metric system is sanctioned by law.	

Foreign.	*U. States.* Winchester bushels.
MOROCCO.—	
MOZAMBIQUE.—Portuguese measures are used here.	
NORWAY.—Same as in Denmark.	
PERSIA.—8 sextarios = 2 chenicas = 1 capicha.	
25 capichas = 8 colothuns = 1 artaba = 21 maunds tabree (customs), or 144 av. lbs., - =	1.8541
22 sextarios = 1 sabbitha, - - - =	0.20395
15 capichas = 1 legana, - - - =	1.11246
PORTUGAL.—16 quartos = 4 alqueires = 1 fanga = 120 arratels weight, - - - =	1.56418
15 fangas = 1 moio = 1800 arratels weight, - =	23.46267
Oporto: 1 fanga = 1¼ fanga of Lisbon, or 150 arratels weight, - - - - =	1.95522
ROME.—88 quartucci = 22 scorzi = 16 starelli = 12 staja = 8 quarterella = 4 quarte = 2 rubbiatilli = 1 rubbio, - - - - =	8.3562
RUSSIA.—32 garnetz = 16 tschetwertka = 4 tschetwerik = 2 payak = 1 osmin, - - =	2.97607
2 osmins = 1 tschetwert; 1½ t. = 1 kuhl.	
SANDWICH ISLANDS.—Same as in the United States.	
SPAIN (*Standard of Castile*).—16 racion = 4 quartillos = 1 celemin, *or* almuda; 12 celamins = 4 cuartilla = 1 fanega = 4⅞ arrobas weight of distilled water at maximum density, or 123.64393 av. lbs., - - - =	1.59277
12 fanegas = 1 cahiz, - - - - =	19.113241
Special and local:	
Alicante: 4 celamins = 1 barcella; 12 b. = 1 cahiz = $\frac{4}{11}$ cahiz of Castile, - - - =	6.950306
Barcelona: 48 picotin = 12 cortan = 1 quartera = 6⅝ Castilian arrobas weight, - - =	2.08285
2½ quarteras = 1 carga; 4 quarteras = 1 salma.	
SWEDEN.—4 quarter = 2 stop = 1 kanna = $\frac{1}{10}$ cubic fot.	
7 kanna = 4 kappe = 1 fjerding, - - =	0.519847
4 fjerding = 1 spann; 2 s. = 1 tunna, - - =	4.158777
36 kappe = 1 tunne (firm measure), - - =	4.678624
SWITZERLAND.—(Official and legal for the 22 Cantons):	
100 immi = 10 viertel, gelt, *or* quarteron = 1 malter = 150 litres, - - - - =	4.25661
Also, 16 mässli = 4 vierling = 1 viertel.	

Foreign.	*U. States.* Winchester bushels.
TRIPOLI. = 2 nufs-orbah = 1 orbah; 4 orbahs = 1 temen; 4 t. = 1 ueba = 216 rotls weight, - =	3.05279
TUNIS. — 12 zah, *or* saha = 1 quiba; 16 q. = 1 caffiso = 425 okas weight, - - - =	15.01663
TURKEY. — 4 kiloz = 1 fortin = 110 okas weight, =	3.986315

West Indies.

Cuba, Porto Rico, San Domingo: Same as in Spain, standard of Castile.

Hayti, Empire of. — 16 litrons = 1 boisseau; 12 b. = 1 setiere, - - - - - = 4.4299

Gaudeloupe, Martinique, Marie Galante, Desirade, northern portion of *St. Martin, Les Saints.* — Official, as in France; also, as in Hayti; and the U. S. bushel is often used.

Jamaica, Trinidad, Bahamas, Barbadoes, St. Christopher, Dominica, Montserrat, Grenada, St. Lucia, Antigua, Tortola, Tobago, Nevis, Virgin Gorda, Grenadines. — Official, as in Great Britain; in trade, generally as in the United States, but in *Trinidad* often as in *Hayti.*

St. Thomas, Santa Cruz, St. Jan. — Official, as in Denmark.

St. Eustatius, Curacoa, Buen Ayre, Oruba, southern portion of *St. Martin.* — Official, as in Holland.

Margarita, Tortuga, Blanquilla. — Same as in Venezuela.

St. Bartholomew. — Official, as in Sweden.

5

MEMORANDA AND ADDENDA,

RELATIVE TO FOREIGN MONEYS OF ACCOUNT, COINS, WEIGHTS, MEASURES, QUOTATIONS OF STOCKS, ETC.

GREAT BRITAIN. — In Great Britain, *sovereigns* weighing not less than 122¾ grains are a legal tender for a pound sterling each; which makes the minimum value of a pound sterling in gold equal to $4.8458226; and this value of the pound sterling, very nearly, is adopted by the United-States Government in assessing duties on British invoices.

Goods by weight, passing through a British custom-house, and subject to duties, are subject to an allowance, called *draft* or *tret*, for supposed waste over and above the actual *tares*, as follows; viz., on 1 cwt. (112 lbs.), 1 lb.; above 1 cwt. and under 2 cwt., 2 lbs.; on 2 cwt. and under 3 cwt., 3 lbs.; on 3 cwt. and under 10 cwt., 4 lbs; on 10 cwt. and under 18 cwt., 7 lbs; and on 18 cwt. and upwards, 9 lbs. These allowances were also made at the United-States custom-houses, until July 14, 1862, when the discontinuance of the practice was ordered by law. These are the chief reasons why goods by weight from the United States fall short of weight at the British custom-houses.

Consols, or Consolidated Annuities, represent a considerable portion of the public debt of Great Britain: they bear interest at the rate of three per cent. a year, payable semi-annually, and are transferable.

The quotations in London of the prices of United-States Bonds, and of American Stocks generally, are in cents per dollar, payable in United-States gold coins; and upon the old basis of $4⅘ to the £. To these quotations, therefore, 9½ per cent. must be added to express the real price.

The silver dollars coined by the British government for circulation in China weigh 415.4 grains, and are $\frac{9}{10}$ fine; they are, therefore, intrinsically of the same worth as Mexican dollars.

FRANCE.—The prices of United-States Bonds, and of American stocks generally, are quoted in Paris in cents per dollar, payable in French silver coins, and upon the conventional Bourse rate of 5 francs to the dollar; whereby the rate of exchange affects the prices. From the Paris quotations, therefore, 2.8172 per cent. must be deducted to express the true prices, when exchange is at par, or when a dollar in United-States gold is quoted at *fr.* 5.14086.

A tolerance of weight in excess of the standard, and in excess only, is allowed in France; while in the United States and in Great Britain, strict conformity to the standards is required; thus it happens that the model standards of weight, sent abroad from France, are commonly found to be slightly in excess of the true standard.

The tolerance or remedy spoken of is as follows:

On *iron weights* of 50 kilogrammes, tolerance 20 grammes; 20 kilogrammes, tolerance 10 grammes; 5 kilogrammes, tolerance 4 grammes; 1 kilogramme, tolerance 1 gramme.

On *copper* or *brass weights* of 20 kilogrammes, tolerance 1½ grammes; 5 kilogrammes, tolerance ½ gramme; 2 kilogrammes, tolerance ¼ gramme; 1 kilogramme, tolerance 1½ decigrammes.

FRANKFORT.—At Frankfort the prices of United-States Bonds, and of American stocks generally, are quoted in cents per dollar, payable either in United States gold coins or in guilders at the conventional rate of 2½ guilders to the dollar. To these quotations, therefore, when payable in guilders, 2.674 per cent. must be added to express the true prices.

SICILY.—Although the French franc is now the official measure of value in Sicily, yet in commercial transactions the old nomenclature and values are still chiefly used, viz: 3600 picioli = 600 grani = 30 tari = 1 oncia = 1 oncetta of Naples = 57.636 grains of fine gold = $2.48217.

UNITED STATES.—Silver dollar-pieces of the standard of 1834 to 1853, and no others, are still coined by this government in greater or less numbers every year; but they are intended mainly for foreign circulation, and bear a premium at home. They are worth, measured by the present standard for silver coins of less denominations than one dollar, $1.0742 each; and measured by the prevailing commercial relations of gold to silver as 15⅝ to 1 for equal weights, they are worth $1.04 each in United-States gold coins of the present standard. See page 2, SEC. A.

CANADA.—The new silver coins of the *dollar* denomination, first coined by the British government in 1870 for circulation in the Dominion of Canada, weigh at the rate of 360 grains to the dollar, and are $\frac{37}{40}$ fine. They are therefore worth 96$\frac{17}{48}$ cents per nominal dollar's worth, measured by the standard silver coins of the United States of less denominations than one dollar; or they are worth $0.93275 per nominal dollar's worth, measured by United-States gold coins. Five of the colonial silver shillings of Canada are equal in value to four quarter-dollars or two half-dollars of the new coin. See CANADA, Dominion of, page 5, SEC. A.

CUSTOMS TARES,

OR TARES AS ALLOWED BY THE UNITED-STATES GOVERNMENT ON DUTIABLE GOODS.

By an Act of Congress passed July 14, 1862, it is provided that when the original invoice is produced at the time of making entry thereof, and the tare shall be specified therein, it shall be lawful for the collector, if he shall see fit, with the consent of the consignees, to estimate the said tare according to such invoice; and that in all other cases the real tare shall be allowed, and may be ascertained, under such regulations as the Secretary of the Treasury may from time to time prescribe; and that hereafter there shall be no allowance for draft. And, in accordance with these provisions, the following rates of tare were adopted: —

	Per cent.
Almonds, in bags	2
" in bales	2½
" in frails	8
Alum, in casks	10
" coarse or ground, in sacks, 2 lbs. per sack.	
Barytes	3
Cheese, in casks or tubs	10
Cassia, in mats	9
Cinnamon, in bales	6
Chicory, in bags	2
Cocoa, in bags	2
" in ceroons	8
Coffee, Rio, in single bags	1
" " in double bags	2
" all others, actual tare.	
Copperas, in casks	10
Currants, in casks	10
Hemp, Manilla, in bales, 4 pounds per bale.	
Hemp, Hamburg, Leghorn, or Trieste, in bales, 5 lbs. per bale.	
Indigo, in ceroons	10
Melado	11
Nails, in bags	2
" in casks	8
Ochre, dry, in casks	8
" in oil, in casks	12
Paris white, in casks	10
Peruvian bark, in ceroons	10
Pepper, in single bags	2
" in double bags	4
Pimento, in bags	2
Raisins, in casks	12
" in boxes	25
" in half-boxes	27
" in quarter-boxes	29
" in frails	4
Rice, in bags	2
Spanish brown, dry, in casks	10
Spanish brown, in oil, in casks	12
Sugar, in mats	2½
" in bags	2
" in barrels	10
" in tierces	12
" in hhds.	12½
" in boxes	14
Salt, fine, in sacks, 3 lbs. per sack.	
Tea, China or Japan, invoice tare.	
Tea, all others, actual tare.	
Tobacco, leaf, in bales, 10 pounds per bale.	
Tobacco, leaf, in bales, extra covers, 12 pounds per bale.	
Whiting, common, in casks	10

SECTION B.

REDUCTIONS, EXCHANGE, INVESTMENTS, MIXED NEGOTIATIONS, &c., &c.

PROPOSITION 1. — When a dollar in gold is worth a dollar and thirty cents in currency, what is the value of a currency dollar?

$1 \div 1.30 = \$0.769231$, or 76 cents 9.231 mills. *Ans.*

PROP. 2. — When gold is nominally at a premium of $35\frac{3}{4}$ per cent., at what discount is currency?

$100 - \frac{100}{1.3575} = 26.335175$ per cent. *Ans.*

PROP. 3. — What is the difference between a given principal, P, and a *deduction* of r per cent. of it?

$P \sim Pr = P(1-r)$, the present worth. *Ans.*

PROP. 4. — What is the difference between a given principal, P, and a *discount* of r per cent. of it?

$$P \sim \frac{Pr}{1+r} = \frac{P}{1+r}, \text{ the present worth. } \textit{Ans.}$$

PROP. 5. — Express the difference per dollar between *deduction* at r per cent., and *discount* at the same rate per cent.

$$(1-r) \sim \frac{1}{1+r}. \quad \textit{Ans.}$$

PROP. 6. — Express the discount, d, on a given principal, P, for a given time in days, t, at a given rate, r, per cent. per annum.

$d = \frac{Ptr}{365+tr} = \frac{Pi}{1+i} = P - w$, i being the interest on 1 dollar for the given time at the given rate, and w the present worth for the same time and rate; whence, $w = \frac{365P}{365+tr} = \frac{P}{1+i}$. $= P - d$. See DISCOUNT, page 129.

PROP. 7. — The foregoing proposition (Prop. 6); except that the time, T, is in years, or years and decimal parts of a year?

$d = PTr \div (1 + Tr)$. *Ans.*

PROP. 8. — The foregoing proposition (Prop. 6), except that the time, m, is given in months, or months and decimal parts of a month?

$d = Pmr \div (12 + mr)$. *Ans.*

PROP. 9. — To convert time in days into calendar months and decimal parts of a month; $m = 12d \div 365$; but the converse of this, viz., $d = 365m \div 12$, is not liable to be true in practice, since different calendar months are made up of an unlike number of days; and, therefore, those immediately under consideration may contain a greater or less number of days.

PROP. 10.—What is the intrinsic equivalent in Federal money of £712 10*s.* 9½*d.* sterling?

$$\left(712 + \frac{12 \times s + d}{12 \times 20}\right) \times 4.86656 = \$3467.6166. \quad \textit{Ans.}$$

PROP. 11.—Reduce $3467.6166 to its intrinsic equivalent in sterling money.

$$\begin{array}{rl} 3467.6166 \div 4.86656 = & 712.53958 \\ & 20 \\ \hline & 10.7916 \\ & 12 \\ \hline & 9.4992 \quad £712\ 10s.\ 9\tfrac{1}{2}d. \quad \textit{Ans.} \end{array}$$

PROP. 12.—Express the equivalent in Federal money, $, of a given quantity in sterling money, £, at a given true or direct rate of exchange, *r*, that is to say, at a rate of exchange based upon the intrinsic equivalent of the two denominations of money.

$$\$ = £ \times 4.86656 \times r. \quad \textit{Ans.}$$

PROP. 13.—Express the equivalent in Federal money of a given quantity in sterling money, rated by the former intrinsic par of 4*s.* 6*d.* sterling to the dollar, the given rate of exchange, *f*, being upon the same basis.

$$\$ = \frac{£ \times 40 \times f}{9} = \frac{£ \times 4.86656 \times f}{1.09498}. \quad \textit{Ans.}$$

EXAMPLE.—What is the equivalent in Federal money of £435 7*s.* 6*d.* sterling, rated by the former intrinsic par of $4⅘ to the £, the rate of exchange upon the same basis being 1.10¼?

$$\frac{435.375 \times 40 \times 1.1025}{9} = \frac{435.375 \times 4.86656 \times 1.1025}{1.09498} =$$

$2133.3375. *Ans.*

PROP. 14.—Reduce F1172.36 (1172 francs, 36 centimes of France) to its equivalent in Federal money?

$$1172.36 \times 0.19452 = 1172.36 \div 5.14086 = \$228.05. \quad \textit{Ans.}$$

PROP. 15.—Express the intrinsic or par equivalent of francs, *F*, or francs and decimal parts of a franc, in Federal money, $; and *vice versa.*

$$\$ = F \times 0.19452 = F \div 5.14086. \quad \textit{Ans.}$$
$$F = \$ \div 0.19452 = \$ \times 5.14086. \quad \textit{Ans.}$$

PROP. 16.—Express the intrinsic equivalent of marks banco, *M*, or marks and decimal parts of a mark banco of Hamburg, London rate, in Federal money; and *vice versa.*

$$\$ = M \times 0.35393 = M \div 2.8254. \quad \textit{Ans.}$$
$$M = \$ \times 2.8254 = \$ \div 0.35393. \quad \textit{Ans.}$$

EXAMPLE.—What is the equivalent in Federal money of 6472 marks, 12 schillings banco of Hamburg, London rate?

$6472.75 \times 0.35393 = 6472.75 + 2.8254 = \$2290.90.$ *Ans.*

PROP. 17.—Which is the most advantageous purchase, other things being equal; viz., bills of exchange on London at $1.10\frac{1}{2}$, rate of 4*s.* 6*d.* sterling to the dollar; or on Paris at F5.15 to the dollar; or on Hamburg at $35\frac{5}{8}$ cents per mark banco?

$1.105 \div 1.09498 = 1.009151-$, London.
$5.15 \div 5.14086 = 1.00178-$, Paris.
$35.625 \div 35.393 = 1.00655+$, Hamburg.
Bills on Paris. *Ans.*

NOTE.—London allows, as the intrinsic par, £1 for F25.21 = 19.304 Federal cents per franc, instead of 19.452, the prevailing commercial par. But at the same time Great Britain values the silver franc of France, measured by the silver in her own silver coinage, at £1 = $\frac{1614.54545+}{69.445575}$ = 23.249+ francs = 20.93234+ Federal cents per franc.

Of Notes for discount, and their avails at bank.

PROP. 18.—When the time of the note is expressly written in days; or when the actual number of days in the specified time are employed; that is, when the *true* time is taken, and taken in days, $P = \frac{365a}{365 - tr}$, and $a = \frac{P(365 - tr)}{365}$, P being the principal or face of the note, a the avails or sum advanced by the bank, t the time including grace, and r the rate of the interest or discount per annum. See BANK DISCOUNT, p. 127.

EXAMPLE.—What must be the principal of a note payable in 90 days, in order that the *equitable* avails at bank, for the time of the note and 3 days grace, the rate being 6 per cent., shall be $1000?

$$\frac{365 \times 1000}{365 - 93 \times .06} = \$1015.53. \quad \textit{Ans.}$$

PROP. 19.—When the time of the note is expressly written in months, and a month is taken to be the $\frac{1}{12}$ part of a calendar year,

$P = \frac{12a}{12 - r(m + 0.1)}$, and $a = \frac{P[12 - r(m \times .1)]}{12}$; m being the time in months, and .1 being $\frac{3}{30}$ of a month, or the usual 3 days grace.

EXAMPLE.—$1000 are to be obtained from a bank, on a note having three months to maturity, and three days grace; the rate being 6 per cent., for what sum must the note be drawn?

$$\frac{12 \times 100}{12 - 3.1 \times .06} = 1015.74. \quad \textit{Ans.}$$

NOTE.—3 calendar months and 3 days, mean time, are more than 93 days, by $1\frac{1}{4}$ day.

PROP. 20. — When the true time of the note is taken in months or days, and a year is assumed to consist of 12 months of 30 days each, or of 360 days only,

$$P = \frac{360a}{360 - tr}, \text{ and } a = \frac{P(360 - tr)}{360}.$$

EXAMPLE. — What must be the principal of a note for discount, payable in 90 days after acceptance, that the proceeds at bank shall be $1000, the rate being 6 per cent., the grace 3 days, and the bank assuming that a year consists of 360 days only?

$$\frac{360 \times 1000}{360 - 93 \times .06} = \$1015.74. \quad \textit{Ans.}$$

NOTE. — The Government of the United States, and the Courts, in matters of interest and discount, reckon time at 365 days to the year; and in Great Britain, France, and all Europe, a year, for like purposes, contains 365 days.

PROP. 21. — A note on time, without interest, dated Jan. 2, 1868, is to be given in exchange for the following obligations; the time of the note is required.

Note, due	March 3, 1868,	for	$370.
Bond, "	April 16,	"	830.50
Note, "	June 11,	"	1120.
Acc't, "	June 4,	"	127.50

From Jan. 2 to March 3 is 61 days × 370 = 22570
" " 2 to April 16 is 105 " × 831 = 87255
" " 2 to June 11 is 161 " × 1120 = 180320
" " 2 to " 4 is 154 " × 127 = 19558
2448)309703(127 days after Jan. 2 = May 8, 1868. *Ans.*

But this is simply equivalent to finding the common time of maturity of the obligations to be transferred, which must be the answer.

EXAMPLE. — Due March 3, $370 × 0 = 0
" April 16, 831 × 44 = 36564
" June 11, 1120 × 100 = 112000
" " 4, 127 × 93 = 11811
2448) 160375(66 days after March 3, = May 8, 1868. See EQUATION OF PAYMENTS, page 132.

PROP. 22. — A note dated March 1, 1869, and bearing interest at 7 per cent. from date, is to be given in exchange for the following obligations; the principal of the note is required.

1869, Feb. 16, settlement note on interest, value of, this day,	$327.36
" March 27, acceptance of this date for	1000.00
" May 4, account, averaging due this date,	658.73

1869, Feb. 16, due \$327
" M'ch. 27, " 1000 × 39 = 39000
" May 4, " 659 × 77 = 50743
\$1986)89743(45.2 days later than Feb. 16, = April 3, 1869, the day on which the given obligations collectively become due, or average due; which is 32 days *later* or *after* March 1, 1869: then, since the adjustment involves discount instead of interest,

$$P = \frac{365a}{365 + tr} = \frac{365 \times 1986}{365 + 32 \times .07} = \$1973.89\text{–}. \quad \textit{Ans.}$$

In which P represents the principal, or face of the new note, a the sum of the obligations to be transferred or cancelled, t the time in days of the adjusting interest or discount, and r the rate of the interest or discount.

PROP. 23. — The foregoing proposition (Prop. 22), except that the new note is to be dated July 1, 1869, instead of March 1, 1869.

From April 3, '69, to July 1, '69, is 59 days, the number of days that April 3d is *earlier* or *previous* to July 1st: then, since the adjustment involves interest instead of discount,

$$P = \frac{a(365 + tr)}{365} = \frac{369.13 \times 1986}{365} = \$2008.47. \quad \textit{Ans.}$$

PROP. 24. — Suppose we equate the time of the following demands by the common rule, and then make up the account correctly by interest and discount to the given dates, and thence to the equated time, by way of exhibiting general principles, and the bearing of the common rule of average upon those principles.

Interest and discount at 7 per cent.

1868, note due March 10 for \$500
" " " June 8 " 750 × 90 = 67500
" " " Sept. 6 " 1050 × 180 = 189000
2300) 256500 (112 days later than March 10, = June 30, '68, the day on which the given demands (\$2300) are assumed to collectively mature.

\$500 due March 10,	worth March 10				\$500.
750 " June 8,	" " 10	(90 days' disc't.)			737.28
1050 " Sept. 6,	" " 10	(180 " ")			1014.96
					\$2252.24;

worth, June 30 (112 days' interest), \$2300.62. (True time $= 365(P - w) \div wr = 112 - 365(W - P) \div Pr.$)

\$500 due March 10,	worth June 8	(90 days' interest)	\$508.63
750 " June 8,	" " 8		750.00
1050 " Sept. 6,	" " 8	(90 days' discount)	1032.18
			\$2290.81;

worth, June 30 (22 days' interest) $2300.48. (True time $= 365(P - w) \div wr + 90$.)

$500 due March 10,	worth Sept. 6	(180 days' interest)		$517.26
750 " June 8,	" " 6	(90 " ")		762.95
1050 " Sept. 6,	" " 6			1050.00
				$2330.21;

worth, June 30 (68 days' discount), $2300.21. (True time $= 180 - 365(w - P) \div Pr$.)

$500 due March 10,	worth June 30	(112 days' int.)	$510.71
750 " June 8,	" " 30	(22 " ")	753.17
1050 " Sept. 6,	" " 30	(68 " disc't.)	1036.48 =

$2300.36. (True time $= 112 - 365(w - P) \div Pr = 111.184$ days.

NOTE. — On the assumption that simple interest is justly due and payable yearly, it is apparent that no demand should enter the account for average at a present worth more than a year due previous to the maturity of the debt latest due.

PROP. 25. — A man sold two horses for $150 apiece, one at a *profit* of 25 per cent., and the other at a *loss* of 20 per cent.; which was the greater, the profit or the loss on the two sales, and what sum of money expresses the difference?

$150 \div (1 - .20) =$ $187.50, cost of the horse sold at a loss, and $187.50 - 150 =$ $37.50 loss; $150 \div (1 + .25) = 120$, cost of the horse sold at a profit, and $150 - 120 =$ $30 profit; then $37.50 - 30 =$ $7.50 loss greater than the profit. *Ans.*

PROP. 26. — A merchant sold two packages of goods for the same sum of money each; on one of them he cleared 25 per cent., and on the other he lost 25 per cent.: did he gain or lose in the aggregate; and, if either, what per cent.?

$1 \div (1 - .25) = 1\frac{1}{3}$, cost relative to the sum received as 1 of the package sold at a loss; $1 \div (1 + .25) = \frac{8}{10}$, cost relative to the sum received as 1 of the package sold at a profit; then

$$\frac{(1\frac{1}{3} - 1) \sim (1 - \frac{8}{10})}{1\frac{1}{3} + \frac{8}{10}} = \frac{4}{64} = \frac{1}{16} = 6\tfrac{1}{4} \text{ per cent. loss.} \quad \textit{Ans.}$$

PROP. 27. — A and B purchased a farm in company for $8000; A paid $5000 in part payment, and B paid the remainder; they then sold $\frac{1}{3}$ of the farm for $4000; and, to close the copartnership, each took $\frac{1}{2}$ of the remaining $\frac{2}{3}$ to his own private interest: how much, if any, of the undivided cash received for the land they disposed of attaches to B's private interest?

$\frac{3}{8}(4000 - \frac{1}{3}$ of $8000) =$ $500, B's share of the profit on the sale, and $3000 + 500 - \frac{1}{3}$ of $8000 =$ $833⅓. *Ans.*

PROP. 28. — *The sum, S, of two numbers, N and n, given, the difference, d, of their respective factors, C and c, to produce like products given, and the sum of the products, P, given, to find the numbers.*

Let $D =$ the difference of the assumed numbers, and let $m = P \div S$; then

1st step, $\frac{\frac{1}{2}S(m + \frac{1}{2}d)}{m} = N$ | $\frac{\frac{1}{2}P}{n} = C$

$S - N = n$ | $\frac{1}{2}P \div N = c$

$N - n = D$ | $C - c = d'$, and

$d' : d :: D : D'$.

2d step, $\frac{S + D'}{2} = N'$ | $\frac{1}{2}P \div n' = C'$

$S - N' = n'$ | $\frac{1}{2}P \div N' = c'$, $C' - c' = d''$;

$d'' : d :: D' : D''$

3d step, $\frac{S + D''}{2} = N''$ | $\frac{1}{2}P \div n'' = C''$

$S - N'' = n''$ | $\frac{1}{2}P \div N'' = c''$, and $C'' - c'' = d'''$, &c.

NOTE.—By this manner of proceeding, all the elements in propositions of this nature may be approximated to any degree of exactness desired; and the true values will be between those obtained by the first step and those obtained by the second. The greater required number (N) will be less than that obtained by the first step, and greater than that obtained by the second. Often but two steps, and seldom more than three, will be required for ordinary practical purposes. When the difference between any trial C and its corresponding c becomes equal to the given difference (d), the elements in that step will be the exact ones sought, if no error has been committed in the work. It is not necessary, however, to obtain the first trial N by the method here proposed; for any number whatever may be assumed in its stead: but, when thus obtained, it has the advantage, generally, of being near the true number sought, and of being known to be greater than that number.

EXAMPLE.—A certain farm containing 80 acres is worth in the aggregate \$62.50 per acre; but one section of it, to the extent of half the gross value, is worth \$11 per acre more than the other; how many acres are there in the lesser section?

$P = 80 \times 62.50 = \$5000$, the gross value of the farm; then

$\frac{(62.5 + 5.5)\,40}{62.5} = 43.52 = N$ | $2500 \div 36.48 = 68.5307+ = C$

$80 - 43.52 = 36.48 = n$ | $2500 \div 43.52 = 57.444853- = c$

$7.04 = D$ | $11.085846 = d'$

$11.085846 : 11 :: 7.04 : 6.985484+ = D'$

$\frac{80 + 6.985484}{2} = 43.492742 = N'$

$80 - 43.492742 = 36.507258 = n'$

$2500 \div 36.507258 = 68.479534 = C'$

$2500 \div 43.492742 = 57.480855 = c$

$10.998679 = d'' =$ 11 nearly; the lesser section, therefore, contains 36.507258 acres, nearly. *Ans.*

NOTE. — All propositions of this general class having given relations of parts contain the requisite elements for direct solutions, and need not be worked, as they commonly are, by rules denominated *Double Position*. For example: "The old sea-serpent's head is 10 feet long, his tail is as long as his head and half the length of his body, and his body is as long as his head and tail; what is the whole length of the monster?"

$h=10$, $t=h+\frac{1}{2}b$, and $b=2h+\frac{1}{2}b$; then

$b-\frac{1}{2}b=2h$, therefore $b=4h=40$

$t=h+\frac{1}{2}b$, therefore $t=10+20=30$

$h=10$. 80 feet. *Ans.*

PROP. 29. — Divide \$1000 into four such parts that the second shall contain \$10 more than the first, the third \$6 more than the second, and the fourth $2\frac{1}{2}$ times as many as the first and second.

Let x represent the smallest part; then

$1000=x+x+10+x+16+5x+25$, and

$1000-(10+16+25)=949=8x$, and $949\div 8=x=$ \$118.625, 1st; \$128.625, 2d; \$134.625, 3d; \$618.125, 4th.

PROP. 30. — A gentleman being asked his own age and the age of his wife, replied, If you subtract 5 years from my age, and divide the remainder by 8, the quotient will be $\frac{1}{3}$ of my wife's age; and if you add 2 years to her age, then multiply the sum by 3, and subtract 7 from the product, the remainder will be my age. Required the age of each.

Let $x=$ the wife's age; then $3x+6-7=$ the husband's age; but $\frac{3x+6-7-5}{8}=\frac{x}{3}$; therefore, $9x+18-36=8x$, and $x=36-18=18$, the wife's age, and $18\times 3+6-7=53$, the husband's age. *Ans.*

PROP. 31. — *Two relations between two unknown numbers given, to find the numbers.*

BY DOUBLE POSITION. — *Special cases.*

Let $x=$ the lesser of the correct numbers sought.

" $N=$ the greater and n the lesser of the assumed values of x, or trial numbers.

" $D=$ the greater and d the lesser of the differences between the trial numbers and those obtained in their stead; then

When both differences are in excess, or show that both the assumed values of x are too high,

$$x=\frac{(N+n)d}{D-d}.$$

When both differences are in deficiency, or show that both the assumed values of x are too low,

$$x=\frac{(N+n)D}{D+d}.$$

When one of the differences is in excess and the other in deficiency, or when one of the assumed values of x is too high and the other too low,

$$x = \frac{ND + nd - D - d}{D + d}.$$

EXAMPLE. — Assume the wife's age (last preceding proposition) is 25, then $(25 + 2) \times 3 - 7 = 74$, the husband's age, and $\frac{(74 - 5) \times 3}{8} = 25.875$, instead of 25, and showing a difference of .875 in excess. Again, assume the wife's age is 20, then $(20 + 2)3 - 7 = 59$, the husband's age, and $\frac{(59 - 5)3}{8} = 20.25$, instead of 20, and showing a difference of .25 in excess; then

$\frac{(25 + 20) \times .25}{.875 - .25} = 18$, the wife's age, and

$(18 + 2) \times 3 - 7 = 53$, the husband's age.

PROOF. $\frac{(53 - 5)3}{8} = 18$.

NOTE. — It is not necessary to assume the value of the lesser required number instead of that of the greater, for the value of either may be assumed as preferred; and when the greater is assumed, the foregoing formulas are not applicable, but a set that are can be easily made.

PROP. 32. — A said to B, Give me one of your apples and I shall then have as many as you will have left. B replied, Give me one of yours, and I shall then have twice as many as you will have left. How many apples had each?

By the first proposition, $A + 1 = B - 1$; therefore $B = A + 2$: but by the second proposition, $B + 1 = 2(A - 1)$; therefore $B = 2A - 3$: then $2A = A + 2 + 3$, and $A = 2 + 3 = 5$; $B = A + 2 = 7$. *Ans.*

PROP. 33. — The lesser and half the greater of two casks of wine = 82 gallons; and the greater and $\frac{1}{3}$ the lesser = 129 gallons: how many gallons are in each cask?

Let x = the greater and y the lesser; then
$x + \frac{1}{3}y = 129$, and $y + \frac{1}{2}x = 82$: therefore
$2y = 164 - x$, and $\frac{1}{3}y = 129 - x$; consequently
$2y - \frac{1}{3}y = 164 - 129$, and $y = 21$; also
$129 - \frac{1}{3}y = x = 129 - 7 = 122$. *Ans.*

PROP. 34. — Smith's several cows are in number to their average cost per head as 45 to 138, and they collectively cost him $690; how many cows has he?

Let n represent the number of cows, and c their average cost; then
$n : c :: 45 : 138$; but $n \times c = 690$, therefore

$$\sqrt{\frac{690 \times 45}{138}} = n = 15. \quad \textit{Ans.}$$

Solved, also, by PROB. X, page 170.

PROP. 35. — An apple-vender bought one-half of a certain lot of apples at the rate of 2 for a cent, and the other half at the rate of 3 for a cent, and concluded that they collectively cost him 2 cents for 5; being willing to dispose of them at cost, he accordingly mixed them together, and sold them out 5 for 2 cents, and lost 5 cents by so doing: how many apples had he?

Let $x =$ the whole number of apples or answer; then

$$x = \left(\frac{\frac{1}{2}x}{2} + \frac{\frac{1}{2}x}{3} - 5\right)\frac{5}{2} = \left(\frac{x}{4} + \frac{x}{6} - 5\right)\frac{5}{2} = \frac{50x - 600}{48}, \text{ therefore}$$

$48\,x + 600 = 50x$, $2x = 600$, and $x = 300$. *Ans.*

NOTE. — The sum of $\frac{1}{2}$ of a quantity and of $\frac{1}{3}$ of a like quantity, is more than $\frac{2}{5}$ of the sum of the two quantities, by $\frac{1}{30}$ of one of the quantities.

PROP. 36. — If a body were to start suddenly into motion, and move 80 miles in the first hour, and in each succeeding hour were to move through three-fourths as much space as in the hour last preceding, and were thus to continue in motion *forever*, what space would it describe?

$$80 \div \tfrac{1}{4} = 320 \text{ miles. } \textit{Ans.}$$

NOTE. — This is simply a question in Geometrical Progression descending, in which the greater extreme is 80, the ratio $\frac{3}{4}$, the less extreme 0, and the sum of the terms is required. The formula, since the ratio is less than unity, becomes $S = \frac{(E-e)\,(1-r)}{r-e} + E = \frac{E \times 1 - e}{r-e}$. See GEOMETRICAL PROGRESSION, p. 151.

PROP. 37. — What sum in ready money, D, may I pay for \$10,000 in stocks, P, that are redeemable at par in T, 3 years, and are bearing interest the while at 7 per cent. a year, payable half-yearly, in order that I may realize 6 per cent. simple interest a year on the investment, supposing that the payments of the interest on the stock are to be kept invested at 7 per cent. a year, from their times of maturity till the stock matures?

$$A = P + pT\left(1 + \frac{R(T-1)}{2} + \tfrac{1}{4}R\right) = P\left[1 + rn\left(1 + \frac{r(n-1)}{2}\right)\right],$$

the amount of the stock at the time of its maturity; in which p represents an interest payment, r the rate of the interest per interval between the payments, and n the whole number of the interest payments; and $D = A \div (1 + R'T) = A \div (1 + an)$, in which R' represents the rate of the discount per annum, or rate of the interest on the investment per annum, and a the rate of the discount per interval between the payments; therefore

$$\frac{10000 + 700 \times 3 \times \left(1 + \frac{.07(3-1)}{2} + \frac{.07}{4}\right)}{1.18} =$$

$$\frac{10000 \times \left[1 + .035 \times 6 \times \left(1 + \frac{.035(6-1)}{2}\right)\right]}{1.18} =$$

\$10,409.96. *Ans.*

PROP. 38. — The last preceding proposition, except that my ready money is worth 6 per cent. compound interest yearly?

$$D = A \div (1+R)^3 = 12283.75 \div 1.191016 = \$10,313.67. \quad \textit{Ans.}$$

See ANNUITIES, p. 156; also, p. 125.

PROP. 39. — The preceding proposition (Prop. 37), except that the payments of the interest are to be invested at 6 per cent. a year, from the times they become due, till the stock matures?

$$\frac{10000 \times (1 + .03 \times 6(1 + \frac{.03 \times 5}{2})}{1.18} = \$10,114.41. \quad \textit{Ans.}$$

PROP. 40. — The preceding proposition (Prop. 37), except that the interest on the stock is payable quarterly?

$$\frac{10000 + 700 \times 3 \times (1 + \frac{.07 \times (3-1)}{2} + \frac{3}{8} \times .07)}{1.18} =$$

$$\frac{10000 \times (1 + .0175 \times 12 \times (1 + \frac{.0175 \times 11}{2})}{1.18} =$$

$\$10,427.22.$ *Ans.*

PROP. 41. — The first proposition in this class (Prop. 37), except that my ready money is worth 6 per cent. interest a year, with the interest payable semi-annually?

$$D = \frac{A}{1 + T(R' + \frac{1}{4}R')} = \frac{A}{1 + T(\frac{r(n-1)}{2})} = \frac{12283.75}{1.225} =$$

$\$10,027.55.$ *Ans.*

PROP. 42. — Express the *difference*, per dollar, between the *amount* of a given algebraic principal for a given algebraic time and rate, and the *present worth* of the same principal for the same time and rate, the time being in days.

$$\frac{365 + tr}{365} \sim \frac{365}{365 + tr} = \left(1 + \frac{tr}{365}\right) \sim \frac{1}{1 + \frac{tr}{365}}. \quad \textit{Ans.}$$

EXAMPLE. — What is the difference between the amount of $1,550 for 175 days at 8 per cent. a year, and the present worth of the same sum for the same time and rate?

$$d = \frac{P(365 + tr)}{365} - \frac{365P}{365 + tr} = P\left(\frac{365 + tr}{365} - \frac{365}{365 + tr}\right) =$$

$\$116,708.$ *Ans.*

PROP. 43. — A purchased a bill of goods on six months' credit, amounting to $2,000, with the understanding that he should be allowed 5 per cent. off for ready cash, in whole or part payment: he paid $1,000 ready cash; for what sum ought he to be credited on the bill?

$$1000 \div (1 - .05) = \$1052.63. \quad \textit{Ans.}$$

PROP. 44. — Which is the lower offer, goods at 1.28¾, on 4 months' credit; or the same goods at 1.30, on 6 months' credit; allowing money to be worth 8 per cent. interest a year?

$$\frac{1.28\frac{3}{4}}{1+\frac{4\times .08}{12}}=\frac{1.2875\times 12}{12.32}=1.254+, \text{ the present worth.}$$

$$\frac{1.30}{1+\frac{6\times .08}{12}}=\frac{1.30\times 12}{12.48}=1.25, \text{ the present worth.}$$

The 1.30 terms, slightly. *Ans.*

Conversion of debts not yet due into others of like sums each, and having a common difference of time from maturity to maturity.

S = gross sum to be converted.

T = time from the present to the maturity of the gross sum.

n = number of common substitutes.

s = common sum of the substitutes.

t = common difference of time from maturity to maturity of the substitutes.

t' = assigned time from the present to the maturity of one of the substitutes.

When the common difference of time is to apply to all the common substitutes, and is to be measured from the present,

$$s = S \div n, \text{ and } t = T \div [\tfrac{1}{2}(n+1)].$$

PROP. 45. — An investment of $2,100 having 90 days to maturity is to be substituted by 3 others of like sums each, which are to become due at the close of a common difference of time from the present, and from one to another: the common sum and common difference of the times are required.

2100 ÷ 3 = $700, the common sum of the substitutes; and

90 ÷ 2 = 45 days, the common difference, or common interval.

The three substitutes of $700 each, therefore, are to be made payable, the first at the expiration of 45 days, the second at the expiration of 90 days, and the third at the expiration of 135 days from the present time.

Proof.

700 × 45 = 31500
700 × 90 = 63000
700 × 135 = 94500
21)189000 = 90; or 700 × (45 + 90 + 135) = 2100 × 90.

PROP. 46. — Five notes are to be made for like sums each, and are to become due at the close of equal intervals of time from the present, and from one to another; and these notes are to be given in exchange for the four following obligations; viz., $1600, due in 90 days, $1250.62 due in 80 days, $852.21 due in 57 days, and $1865 due in 175 days, from the present time. The common denomination of the notes, and the common interval of time are required.

1600.00 × 90 = 144000
1250.62 × 80 = 100050
852.21 × 57 = 48576
1865.00 × 175 = 326375
5567.83) 619001 = 111 days, the mean or average time of maturity from the present of the obligations to be converted; and

5567.83 ÷ 5 = \$1113.57, the common denomination of the notes; and

111 ÷ ½(5 + 1) = 37 days, the common difference of time, or common interval.

The special times to maturity, therefore, of the five notes of \$1113.57 each, are 37, 74, 111, 148, 185 days from the present time.

When one of the common substitutes is to be treated as cash, or is to be considered as due at the present time, and the common interval is to be measured from the present,

$$s = S \div n, \text{ and } t = T \div [\tfrac{1}{2}(n - 1)].$$

Prop. 47. — Several matters of indebtedness, amounting in the aggregate to \$2175.44, and which will collectively mature, or become due by average, at the close of 68 days from the present time, are to be cancelled by the payment of one-fourth of their sum down, and by passing three notes, made for one-fourth of their sum each, and payable at the close of a common difference of time from the present and from one to another. The common sum and common difference are required.

2175.44 ÷ 4 = \$543.86, the common sum; and

68 ÷ ½(4 — 1) = 68 ÷ 1.5 = 46 days, the common interval.

The three notes, therefore, are to be made for \$543.86 each, and are to be made payable, the first at 46, the second at 92, and the third at 138 days from the present time.

When the common interval is to apply between all the common substitutes, and is to be measured from an assigned time for the maturity of one of them,

$$s = S \div n, \text{ and } t = (T \sim t') \div [\tfrac{1}{2}(n - 1)].$$

Prop. 48. — A debt of \$4500, due 123 days hence, without interest, is to be substituted by 4 notes, made for one-fourth of the sum each, which are to run an equal interval of time from maturity to maturity, and one of them is to be made payable at the close of an interval of 30 days from the present time. The common sum and common interval are required.

$4500 \div 4 = \$1125$, the common denomination of the notes; and $(123 - 30) \div 1.5 = 62$ days, the common difference of time.

The times to maturity of the substitutes, therefore, are 30, 92, 154, 216 days later than the present time.

PROP. 49. — The last preceding proposition (Prop. 47), except that the given debt of $4500 has but 75 days to maturity; and one of the notes is not to become due until the lapse of 105 days from the present time.

$(75 \sim 105) \div 1.5 = 20$ days, the common interval.

The times from the present to the maturities of the notes, therefore, are 45, 65, 85, 105 days.

PROP. 50. — It is proposed to relinquish obligations, amounting in the aggregate to $2554.72, and which will collectively become due by equation at the close of 40 days from the present time, and to receive in their place 3 notes, made for one-fourth of the sum each, and the balance in ready cash; the said notes to run equal intervals of time from maturity to maturity, measured from the present, and one of them to run 85 days; the adjusting interest, or discount, to be at 7 per cent.

$(85 - 40) \div \frac{1}{2}(4 - 1) = 30$ days, the common interval.

The three notes of $638.68 each, therefore, are to be made payable at 85, 55, 25 days; and the present worth of the cash payment of $638.68, due $30 - 25 = 5$ days hence, is $638.07.

To invest a given sum of money in parts, at unlike rates of interest, and the parts to gain like interest in equal intervals of time.

p, p', p'', &c. = the parts, or partial investments.

$S =$ the sum of the investments.

r, r', r'', &c. = the given rates, or these in their relations to each other, expressed in any proportion preferred.

$m =$ the product of r, r', r'', &c., as expressed.

$N =$ the sum of $\frac{m}{r} + \frac{m}{r'} + \frac{m}{r''}$, &c.

p, p', p'', &c. $= \frac{Sm}{Nr} = \frac{Sm}{Nr'} = \frac{Sm}{Nr''}$, &c., inversely.

PROP. 51. — Twenty thousand dollars ($20,000) are to be placed at interest in three such parts that the interest on them, at their respective rates of 6, 7, and 8 per cent. a year, shall be equal for all like intervals of time. The parts, or special investments, are required.

$m = 6 \times 7 \times 8 = 336$; and $336 \div 6 = 56$
$336 \div 7 = 48$
$336 \div 8 = 42$
146, the value of N; $\therefore$

$$p = \frac{20000 \times 56}{146} = \frac{20000 \times 28}{73} = \$7,671.23$$

$$p' = \frac{20000 \times 24}{73} \quad . \quad . \quad . \quad . \quad . \quad = \quad 6,575.34$$

$$p'' = \frac{20000 \times 21}{73} \quad . \quad . \quad . \quad . \quad . \quad = \quad 5,753.43$$

$$\$20,000.00$$

To invest a given sum of money in parts, at like rates of interest, and for unequal intervals of time; and the parts to gain like interest at the close of their respective times.

t, t', t'', &c. = the given times, or these in their relations to each other, expressed in whatever proportion preferred.

m = product of t, t', t'', &c., as expressed.

N = sum of $\frac{m}{t} + \frac{m}{t'} + \frac{m}{t''}$, &c.

S, and p, p', p'', &c., as in the last preceding proposition.

p, p', p'', &c., $= \frac{Sm}{Nt} = \frac{Sm}{Nt'} = \frac{Sm}{Nt''}$, &c., inversely.

PROP. 52. — It is proposed to place $25,000 at interest in four separate sums, one of them for 60, one for 80, one for 110, and one for 150 days' time; and that these sums shall be such, that, at like rates of interest, they will gain like interest at the close of their respective times. The special sums are required.

$m = 60 \times 80 \times 110 \times 150$, or $6 \times 8 \times 11 \times 15$, or $3 \times 4 \times 5.5 \times 7.5 = 495$; and

$$495 \div 3 = 165.$$
$$495 \div 4 = 123.75$$
$$495 \div 5.5 = 90.$$
$$495 \div 7.5 = 66.$$

$444.75 = N$; therefore

$$p = \frac{25,000 \times 165}{444.75} = \frac{25,000 \times 33}{88.95} = \$9,274.87$$

$$p' = \frac{25,000 \times 12375}{44475} = \frac{25,000 \times 495}{1779} = \quad 6,956.16$$

$$p'' = \frac{25,000 \times 90}{444.75} \quad . \quad . \quad . \quad . \quad . \quad . \quad . \quad = \quad 5,059.02$$

$$p''' = \frac{25,000 \times 66}{444.75} \quad . \quad . \quad . \quad . \quad . \quad . \quad . \quad = \quad 3,709.95$$

$$\$25,000.00$$

To invest a given sum of money in parts, at unlike rates of interest, and for unequal intervals of time; and the parts to gain like interest at the close of their respective times.

m = product of the given times, or of their relations to each other, by any measure whatever, that is common to them; or of the given rates, if preferred.

S, N, p, p', p'', &c., as in the preceding.

$$p, p', p'', \text{\&c.} = \frac{Sm}{Ntr} = \frac{Sm}{Nt'r'} = \frac{Sm}{Nt''r''}, \text{\&c.}$$

PROP. 53. — Ten thousand dollars ($10,000) are to be placed at interest in three separate sums, one of them at 4 per cent., for 240 days; one at 6 per cent., for 120 days; and one at 8 per cent., for 80 days; and these sums are to be such that they will gain like interest, one with another, at the close of their respective times. The special sums are required.

$240 \times 120 \times 80$, or $24 \times 12 \times 8$, or $6 \times 3 \times 2 = 36 = t \times t' \times t'' = m.$
4, 6, 8, or 2, 3, 4, $= r, r', r''$
12, 9, 8, the products of tr, $t'r'$, $t''r''$, and

$36 \div 12 = 3$
$36 \div 9 = 4$
$36 \div 8 = 4.5$
$11.5 = N$, therefore

$$p'' = \frac{10{,}000 \times 30}{115} = \frac{10{,}000 \times 6}{23} = \$2{,}608.70$$

$$p' = \frac{10{,}000 \times 40}{115} = \frac{10{,}000 \times 8}{23} = 3{,}478.26$$

$$p = \frac{10{,}000 \times 45}{115} = \frac{10{,}000 \times 9}{23} = 3{,}913.04. \quad \$10{,}000.00.$$

To invest a given sum of money in parts, at like rates of interest, and for unequal intervals of time; and the amount (principal and interest) of the parts to be equal at the close of their respective times.

PROP. 54. — It is proposed to place $16,000 at interest in four separate sums, each at 7 per cent. a year: one of them for 80, one for 100, one for 150, and one for 200 days' time; and that these sums shall be such that their amount shall be equal, one with another, at the close of their respective times. The special sums are required.

$$N = \text{sum of } \frac{365}{365 + rt} + \frac{365}{365 + rt'} + \frac{365}{365 + rt''}, \text{\&c.}, = \frac{1}{1 + \frac{rt}{365}}, \text{\&c.}$$

$$p, p', p'', \&c. = \frac{365S}{N(365+rt)} = \frac{365S}{N(365+rt')} = \frac{365S}{N(365+rt'')},$$

$\&c., = \frac{S}{1+\frac{rt}{365}}$, &c.; then

$365 \div 370.6 = 0.98489$
$365 \div 372. = 0.98118$
$365 \div 375.5 = 0.97204$
$365 \div 379. = 0.96306$
3.90117, the value of N; and

$$p = \frac{5,840,000}{370.6N} = \$4039.36$$
$$p' = \frac{5,840,000}{372N} = 4024.16$$
$$p'' = \frac{5,840,000}{375.5N} = 3986.65$$
$$p''' = \frac{5,840,000}{379N} = 3949.83. \quad \$16,000.00$$

Therefore, $A = S \div N = \$4101.33$, the common amount.

If the times, t, t', t'', &c., be taken in years instead of days, then $N = \frac{1}{1+rt} + \frac{1}{1+rt'} + \frac{1}{1+rt''}$, &c.; and $p = S \div N(1+rt)$; $p' = S \div N(1+rt')$, &c. And, if the times be taken in months, $N = \frac{12}{12+rt} + \frac{12}{12+rt'}$, &c.; and $p = 12S \div N(12+rt)$; $p' = 12S \div N(12+rt')$, &c.

To invest a given sum of money in parts, at unlike rates of interest, and for unequal intervals of time; and the amount of the parts to be equal, one with another, at the close of their respective times.

PROP. 55. — The last preceding proposition, except that the rates are to be 6 per cent. for the 80 days' term, 7 per cent. for the 100 days' term, 8 per cent. for the 150 days' term, and 9 per cent. for the 200 days' term, instead of 7 per cent. for each of the terms.

$365 \div 369.8 = 0.98702$
$365 \div 372. = 0.98118$
$365 \div 377. = 0.96817$
$365 \div 383. = 0.95300$
3.88937, the value of N; and

$p = 5,840,000 \div 369.8N = \$4060.38$
$p' = 5,840,000 \div 372N = 4036.37$
$p'' = 5,840,000 \div 377N = 3982.83$
$p''' = 5,840,000 \div 383N = 3920.44. \quad \$16,000.02.$

To solve Problems in Medial Proportion by the common rule of Simple Proportion; making use of two or more equations when there are three or more given extreme rates, of one or more assumed mean rates when there are three or more given extreme rates, and using the first assumed mean rate as an extreme rate in the second equation, the second assumed mean rate as an extreme rate in the third equation, &c.

Let a represent the higher, c the lower, and b the mean rate, employed in each equation; also, let n represent the sum of the proportional terms already found, or quantity taken at one of the rates, and let x represent the required proportional term, or quantity to be taken at the other rate, in the same equation; then

When the quantity taken at the higher rate in the equation is given,

$x = \frac{(a-b)n}{b-c}$; or, The difference of the lower rate and mean rate, is to the difference of the higher rate and mean rate, as the quantity taken at the higher rate, is to the quantity required, and to be taken at the lower rate. Conversely—

When the quantity taken at the lower rate in the equation is given.

$x = \frac{(b-c)n}{a-b}$; or, The difference of the higher rate and mean rate, is to the difference of the lower rate and mean rate, as the quantity taken at the lower rate, is to the quantity required, and to be taken at the higher rate.

From the foregoing it will be perceived that the initial proportional term, or first quantity, n, may be taken at any number whatever; but, commonly, it will be best to take it at 1.

The assumed mean rates may be taken at any stage between their respective extremes; and, by varying them, an almost endless number of different proportions may be obtained. Even when there are only three extreme rates in the given proposition, the assumed mean rate can commonly be so taken as to obtain the relation of the terms, one to another, that may be desired.

PROP. 56.—In what proportion must two kinds of tea, one rated at 80 cents a pound, and the other at 96 cents a pound, be taken, that the mean rate, or rate of the sum of the quantities taken, shall be 85 cents a pound?

Let 1 represent the initial proportional term, or quantity taken of the kind rated at 96 cents a pound; then

$\frac{(96-85)\times 1}{85-80} = 2\frac{1}{5}$, the required corresponding term, or quantity to be taken of the kind rated at 80 cents a pound. The two kinds of tea, therefore, must be taken in the proportion of one pound at 96 cents a pound, to $2\frac{1}{5}$ pounds at 80 cents a pound, that the mean rate, or rate of the mixture, shall be at 85 cents a pound; or, they must be taken in the proportion of $1 \times 5 = 5$ pounds at 96 cents a pound, to $2\frac{1}{5} \times 5 = 11$ pounds at 80 cents a pound, &c.

Conversely.—How much tea at 96 cents a pound, must be taken with 1 pound at 80 cents a pound, that the mean rate, or rate of the mixture, shall be at 85 cents a pound?

$\frac{(85-80) \times 1}{96-85} = \frac{5}{11}$ of a pound. *Ans.* The two kinds of tea, therefore, taken in the proportion of 1 pound at 80 cents a pound, to $\frac{5}{11}$ of a pound at 96 cents a pound, or in the proportion of 1 times 11 $= 11$ pounds at 80 cents a pound, to $\frac{5}{11} \times 11 = 5$ pounds at 96 cents a pound, as before, will form a mixture at the rate or price 85 cents a pound. See ALLIGATION, page 140; also, see page 117.

PROP. 57.—In what proportion may oats at 50 cents a bushel, and rye at 100 cents a bushel, be mixed with 1 bushel of corn at 80 cents a bushel, that the mean rate, or rate of the mixture, shall be 75 cents a bushel?

Let the assumed mean rate be 60 cents a bushel; then

$$? \text{ at } 50, 1 \text{ at } 80, \text{m } 60 = \frac{(80-60) \times 1}{60-50} = 2 \text{ bus. of oats,} \quad \left.\begin{matrix}\text{1 bus. of corn,}\\ \text{2 bus. of oats,}\end{matrix}\right\} \text{making 3 bus. at 60 cents a bus., and}$$

$$? \text{ at } 100,\ 3 \text{ at } 60, \text{ m } 75 = \frac{(75-60) \times 3}{100-75} = 1\tfrac{4}{5} \text{ bus. of rye.}$$

On the contrary, let the assumed mean rate be 70 cts. a bushel; then

$$\left.\begin{matrix} & \text{1 bus. of corn,} \\ ? \text{ at } 50, 1 \text{ at } 80, \text{ m } 70 = \frac{(80-70) \times 1}{70-50} = \frac{1}{2} & \text{bus. of oats,} \\ ? \text{ at } 100, 1\frac{1}{2} \text{ at } 70, \text{ m } 75 = \frac{(75-70) \times 1\frac{1}{2}}{100-75} = \frac{3}{10} & \text{bus. of rye.} \end{matrix}\right\} Ans.$$

Let the assumed mean rate be 65; then

$$\left.\begin{matrix} & \text{1 bus. at 80 cents,} \\ ? \text{ at } 50, 1 \text{ at } 80, \text{ m } 65 = \frac{(80-65) \times 1}{65-50} = 1 & \text{bus. at 50 cents,} \\ ? \text{ at } 100, 2 \text{ at } 65, \text{ m } 75 = \frac{(75-65) \times 2}{100-75} = \frac{4}{5} & \text{bus. at 100 cents.} \end{matrix}\right\} Ans.$$

PROOF (last example).

```
  1  at  80 = 80
  1  at  50 = 50
 0.8 at 100 = 80
 ---          ---
 2.8        )210 ( 75.
```

PROP. 58.—In what proportion may one quality of wine, rated at \$1.00 a gallon, and another quality, rated at \$1.50 a gallon, be mixed with 1 gallon of water, rated at 0 a gallon, that the mean rate, or average rate of the mixture, shall be \$1.25 a gallon?

Let the assumed mean rate be 25 cents a gallon; then

$$\left.\begin{array}{lr} & \text{1 gal. at } \quad 0 \\ ?\text{ at } \$1.00,\ 1 \text{ at } 0,\ \text{m } 0.25 = \dfrac{(.25-0)\times 1}{1.00-.25} = \tfrac{1}{3} & \text{`` `` } \$1.00 \\ ?\text{ at } \$1.50,\ 1\tfrac{1}{3} \text{ at } .25,\ \text{m } \$1.25 = \dfrac{(1.25-.25)\times 1\tfrac{1}{3}}{1.50-1.25} = 5\tfrac{1}{3} & \text{`` `` } \$1.50 \end{array}\right\} \textit{Ans.}$$

Let the assumed mean rate be 50 cents a gallon; then

$$\left.\begin{array}{lr} & \text{1 gallon at } \quad 0 \\ ?\text{ at } 100,\ 1 \text{ at } 0,\ \text{m } 50 = \dfrac{(50-0)\times 1}{100-50} = 1 & \text{`` `` } \$1.00 \\ ?\text{ at } 150,\ 2 \text{ at } 50,\ \text{m } 125 = \dfrac{(125-50)\times 2}{150-125} = 6 & \text{`` `` } \$1.50 \end{array}\right\} \textit{Ans.}$$

Let the assumed mean rate be 70 cents a gallon; then

$$\left.\begin{array}{lr} & \text{1 gallon at } \quad 0 \\ ?\ 100,\ 1 \text{ at } 0,\ \text{m } 70 = \dfrac{(70-0)\times 1}{100-70} = 2\tfrac{1}{3} & \text{`` `` } \$1.00 \\ ?\ 150,\ 3\tfrac{1}{3} \text{ at } 70,\ \text{m } 125 = \dfrac{(125-70)\times 3\tfrac{1}{3}}{150-125} = 7\tfrac{1}{3} & \text{`` `` } \$1.50 \end{array}\right\} \textit{Ans.}$$

Let the assumed mean rate be 90 cents a gallon; then

$$\left.\begin{array}{lr} & \text{1 gallon at } \quad 0 \\ ?\text{ at } 100,\ 1 \text{ at } 0,\ \text{m } 90 = \dfrac{(90-0)\times 1}{100-90} = 9 & \text{`` `` } \$1.00 \\ ?\text{ at } 150,\ 10 \text{ at } 90,\ \text{m } 125 = \dfrac{(125-90)\times 10}{150-125} = 14 & \text{`` `` } \$1.50 \end{array}\right\} \textit{Ans.}$$

PROP. 59.—In what proportions may four grades of sugar, rated at 8, 10, 14, and 17 cents a pound, respectively, be taken, that the mean rate shall be 12 cents a pound?

Let the assumed mean rates be 9 and 11; then

$$\left.\begin{array}{r} 1 \text{ at } 10 \\ ?\text{ at } 8,\ 1 \text{ at } 10,\ \text{m } 9 = \dfrac{(10-9)\times 1}{9-8} = 1 \text{ at } 8 \\ ?\text{ at } 14,\ 2 \text{ at } 9,\ \text{m } 11 = \dfrac{(11-9)\times 2}{14-11} = 1\tfrac{1}{3} \text{ at } 14 \\ ?\text{ at } 17,\ 3\tfrac{1}{3} \text{ at } 11,\ \text{m } 12 = \dfrac{(12-11)\times 3\tfrac{1}{3}}{17-12} = \tfrac{2}{3} \text{ at } 17 \end{array}\right\} \textit{Ans.}$$

Let the assumed mean rates be $9\frac{1}{2}$ and 11; then

$$\left.\begin{array}{r} 1 \text{ at } 10 = 3 \text{ at } 10 \\ ?\text{ at } 8,\ 1 \text{ at } 10,\ \text{m } 9\tfrac{1}{2} = \dfrac{(10 = 9.5)\times 1}{9.5-8} = \tfrac{1}{3} \text{ at } 8 = 1 \text{ at } 8 \\ ?\text{ at } 14,\ 1\tfrac{1}{3} \text{ at } 9\tfrac{1}{2},\ \text{m } 11 = \dfrac{(11-9\tfrac{1}{2})\times 1\tfrac{1}{3}}{14-11} = \tfrac{2}{3} \text{ at } 14 = 2 \text{ at } 14 \\ ?\text{ at } 17,\ 2 \text{ at } 11,\ \text{m } 12 = \dfrac{(12-11)\times 2}{17-12} = \tfrac{2}{5} \text{ at } 17 = 1\tfrac{1}{5} \text{ at } 17 \end{array}\right\} \textit{Ans.}$$

Let the assumed mean rates be 9 and $10\frac{1}{2}$; then

$$\left.\begin{array}{r} 1 \text{ at } 8 = 7 \text{ at } 8 \\ ? \text{ at } 10, 1 \text{ at } 8, \text{ m } 9 = \dfrac{(9-8)\times 1}{10-9} = 1 \text{ at } 10 = 7 \text{ at } 10 \\ ? \text{ at } 14, 2 \text{ at } 9, \text{ m } 10\frac{1}{2} = \dfrac{(10\frac{1}{2}-9)\times 2}{14-10\frac{1}{2}} = \frac{6}{7} \text{ at } 14 = 6 \text{ at } 14 \\ ? \text{ at } 17, 2\frac{6}{7} \text{ at } 10\frac{1}{2}, \text{ m } 12 = \dfrac{(12-10\frac{1}{2})\times 2\frac{6}{7}}{17-12} = \frac{6}{7} \text{ at } 17 = 6 \text{ at } 17 \end{array}\right\} \textit{Ans.}$$

NOTE.—All the assumed mean rates, it will be perceived, must be taken short of the given, or final, mean rate.

To find the Proportional Coefficients of the dividend, or the numerical relation of each part to the whole, in Partitive Proportion, when the Proportional terms, or terms of the ratio of the parts, one to another, are given.

RULE I.—Divide each of the given proportional terms by the sum of those terms, or by that sum divided by any common measure, or the greatest common measure, of both its parts; the quotients thus obtained, or these reduced to lower or their lowest terms, will be the coefficients demanded; for they will be to one another as their respective proportional terms, which are to one another as the parts of the dividend required; and, collectively taken, they will be equal to a unit, or the whole.

RULE II.—Reduce the given proportional terms to integers, by reducing them to a common denominator, and take the sum of the integers, or numerators of the fractions thus obtained, for the denominator of each integer; the fractions thus formed, or these divided by any common measure of their respective parts, will be the proportional coefficients of the dividend, or coefficients required; for the denominator of any proportional coefficient in partitive proportion, is to its numerator, as the dividend to be divided into parts, is to the part corresponding to that numerator.

It is, therefore, evident that when the proportional terms are given in integers, the integers are to be taken as the numerators, and their sum as the denominator of each integer, to express the numerical relation of each part to the whole, in partitive proportion.

PROP. 60.—It is required to divide a certain quantity into four parts which shall be to one another as $\frac{1}{3}$, $\frac{1}{4}$, $\frac{1}{5}$, $\frac{1}{6}$; that is, the first part to the second as $\frac{1}{3}$ to $\frac{1}{4}$, the second to the third as $\frac{1}{4}$ to $\frac{1}{5}$, and the third to the fourth as $\frac{1}{5}$ to $\frac{1}{6}$; the proportional coefficients of the quantity, or numerical relations of the proposed parts to the whole quantity, are demanded.

By Rule I.—$\frac{1}{3}+\frac{1}{4}+\frac{1}{5}+\frac{1}{6} = \frac{342}{360} = \frac{19}{20}$; and $\frac{1}{3}\div\frac{19}{20} = \frac{20}{57}$, $\frac{1}{4}\div\frac{19}{20} = \frac{20}{76}$, $\frac{1}{5}\div\frac{19}{20} = \frac{20}{95}$, and $\frac{1}{6}\div\frac{19}{20} = \frac{20}{114}$, the coefficients required; or, $\frac{20}{57}$, $\frac{20}{76}$, $\frac{20}{95}$, $\frac{20}{114}$, $= \frac{20}{57}$, $\frac{15}{57}$, $\frac{12}{57}$, $\frac{10}{57} = \frac{20}{57}$, $\frac{5}{19}$, $\frac{4}{19}$, $\frac{10}{57}$, quantities of the same relative value, one to another, as the preceding; and, therefore, the proportional coefficients required, if preferred.

By Rule II.—$\frac{1}{3}$, $\frac{1}{4}$, $\frac{1}{5}$, $\frac{1}{6}$, reduced to fractions having a common denominator, become $\frac{120}{360}$, $\frac{90}{360}$, $\frac{72}{360}$, $\frac{60}{360}$; and by removing the denominators, we obtain the integers 120, 90, 72, 60, quantities bearing the same relation, one to another, as the quantities from which they were derived; and, therefore, since the sum of these integers is equal to 342, we have $\frac{120}{342}$, $\frac{90}{342}$, $\frac{72}{342}$, $\frac{60}{342} = \frac{20}{57}$, $\frac{15}{57}$, $\frac{12}{57}$, $\frac{10}{57} = \frac{20}{57}$, $\frac{5}{19}$, $\frac{4}{19}$, $\frac{10}{57}$, the numerical relations of the proposed parts to the whole, or the proportional coefficients of the dividend required.

Thus, allowing the proportional coefficients to be $\frac{20}{57}$, $\frac{5}{19}$, $\frac{4}{19}$, $\frac{10}{57}$, and the dividend to be \$120; the required parts, in the denomination of the dividend, will be $\frac{120 \times 20}{57} = \$42\frac{2}{19}$, $\frac{120 \times 5}{19} = \$31\frac{11}{19}$, $\frac{120 \times 4}{19} = \$25\frac{5}{19}$, $\frac{120 \times 10}{57} = \$21\frac{1}{19}$.

Proof.—$42 + 31 + 25 + 21 + \frac{19}{19} = 120$. See FELLOWSHIP, page 138.

PROP. 61.—A, B, and C entered into copartnership in a manufacturing business with a gross capital of \$31,000, of which A furnished \$12,500, B \$10,500, and C \$8,000; their collective gain in the business is \$32,800, which is to be divided between the partners in proportion to their respective investments, or share in the gross capital; what is each partner's share of the gain?

The given proportional terms, 12500, 10500, 8000, taken in lower corresponding terms = 125, 105, 80 = 25, 21, 16, the sum of which last set = 62; then $\frac{25}{62}$, $\frac{21}{62}$, $\frac{16}{62} = \frac{25}{62}$, $\frac{21}{62}$, $\frac{8}{31}$, are the proportional coefficients of the dividend for the parts required; therefore,

$$\left.\begin{array}{l} 32{,}800 \times 25 \div 62 = \$13{,}225.81, \text{ A's share,} \\ 32{,}800 \times 21 \div 62 = \$11{,}109.68, \text{ B's share,} \\ 32{,}800 \times 16 \div 62 = \$\ \ 8{,}464.51, \text{ C's share.} \end{array}\right\} \textit{Ans.}$$

PROP. 62.—Two travelers start from two points 80 miles apart, and travel uniformly toward each other, one at the rate of $3\frac{1}{2}$ miles an hour, and the other at the rate of $4\frac{1}{4}$ miles an hour; how many miles will each travel till they meet, supposing the one making $3\frac{1}{2}$ miles an hour starts 3 hours earlier than the other?

$80 - 3 \times 3\frac{1}{2} = 69.5$, and $3\frac{1}{2}$, $4\frac{1}{4} = \frac{7}{2}$, $\frac{17}{4} = \frac{14}{4}$, $\frac{17}{4}$, coef. $= \frac{14..17}{31}$, then

$$\left.\begin{array}{l} 69.5 \times 14 \div 31 + 3 \times 3\frac{1}{2} = 41.8871 \\ 69.5 \times 17 \div 31 \qquad\qquad = 38.1129 \end{array}\right\} \textit{Ans.}$$

PROP. 63.—Two travelers, each under like circumstances making 4 miles an hour, start at the same time from two points 30 miles apart, and travel toward each other, but one, having a down grade, makes $\frac{1}{2}$ mile more per hour in consequence, while the other, having an up grade, makes $\frac{1}{2}$ mile less per hour in consequence; how many miles will each have traveled when they meet?

$4 + \frac{1}{2}$, $4 - \frac{1}{2} = 4\frac{1}{2}$, $3\frac{1}{2}$, and $4\frac{1}{2} + 3\frac{1}{2} = 8$; then

$$\left.\begin{array}{l} 30 \times 4\frac{1}{2} \div 8 = 16\frac{7}{8} \\ 30 \times 3\frac{1}{2} \div 8 = 13\frac{1}{8} \end{array}\right\} \textit{Ans.}$$

When the required parts are given in relation as greater or less, one than another, by a given part of one of them, to find the proportional terms, &c.

RULE.—Let 1 represent the first or initial proportional term in all cases, and let 1 plus the given part of that term greater, or 1 minus the given part of that term less, as the case may be, represent the next proportional term, and so on for all the proportional terms involved in the proposition; then proceed for the proportional coefficients of the dividend, &c., by Rule I or II, as in the foregoing.

PROP. 64.—How shall I divide a certain number of apples between two boys, so that one of them shall receive as many as the other and $\frac{2}{7}$ of as many more? that is, so that the difference of their shares shall be equal to $\frac{2}{7}$ of the smaller share?

$1\,,\,1+\frac{2}{7}=\frac{7\,..\,9}{7}$; coef.$=\frac{7\,..\,9}{7+9}$. Give one of them $\frac{7}{16}$ of the number, and the other $\frac{9}{16}$ of it. *Ans.*

Suppose the number to be divided is 20. Then give one of them $20\times7\div16=8\frac{3}{4}$ apples, and the other $20\times9\div16=11\frac{1}{4}$ apples.

Proof.—$\frac{7}{16}$: $8\frac{3}{4}$ :: $\frac{9}{16}$: $11\frac{1}{4}$, $8\frac{3}{4}+\frac{2}{7}$ of $8\frac{3}{4}=11\frac{1}{4}$, and $8\frac{3}{4}+11\frac{1}{4}=20$.

PROP. 65.—A certain quantity of wheat is to be divided into two parts, such that the difference of the parts shall be equal to $\frac{3}{5}$ of the greater part; the proportional relations of the parts to the whole are required.

$$1\,,\,1-\frac{3}{5}=\frac{5\,..\,2}{5};\ \text{coef.}=\frac{5}{7}\,,\,\frac{2}{7}.\quad \textit{Ans.}$$

PROP. 66.—Divide $1,572 between two persons, so that one shall have $\frac{9}{11}$ as much as the other.

$$1\,,\,1-\frac{2}{11}=\frac{11\,..\,9}{11};\ \text{coef.}=\frac{11\,..\,9}{20},\ \text{then}$$

$\$1{,}572\times11\div20=\$864.60$, and $\$1{,}572\times9\div20=\$707.40$.

PROP. 67.—It is required to divide $5,000 into three parts, such that the second shall be greater than the first by $\frac{1}{6}$ of the first, and the third less than the first by $\frac{1}{4}$ of the first; what are the parts in the denomination of the dividend?

$\overline{1\,,\,1+\frac{1}{6}\,,\,1-\frac{1}{4}}=\frac{6}{6}\,,\,\frac{7}{6}\,,\,\frac{3}{4}$; coef.$=\frac{12\,..\,14\,..\,9}{35}$: 1st $=\$1{,}714\frac{2}{7}$; 2d $=\$2{,}000$; 3d $=\$1{,}285\frac{5}{7}$. *Ans.*

PROP. 68.—D's interest in the ship Hunter is to be transferred to three parties, A, B, and C, in such proportion, one to another, that B's share shall be less than A's by $\frac{1}{4}$ of A's, and C's greater than B's by $\frac{1}{6}$ of B's. The proportional coefficients of the dividend for the respective shares are required.

Let 1 represent the proportional term of A's share; then

$A = 1$, $B = 1 - \frac{1}{4}$, $C = (1 - \frac{1}{4}) + \frac{1 - \frac{1}{4}}{5}$; reduced $= \frac{4}{4}, \frac{3}{4}, \frac{9}{10}$; coef. $= \frac{20..15..18}{53}$.

Let 1 represent the proportional term of B's share; then

$B = 1$, $C = 1 + \frac{1}{5}$, $A = 1 + \frac{1}{3} = \frac{5}{5}, \frac{6}{5}, \frac{4}{3}$; coef. $= \frac{15..18..20}{53}$, as before.

Let 1 represent the proportional term of C's share; then, since by the stipulations C (1) is greater than B by $\frac{1}{5}$ of B, 6 B = 5 and $B = \frac{5}{6}$; also, since by the stipulations B ($\frac{5}{6}$) is less than A by $\frac{1}{4}$ of A, $A - \frac{1}{4}A = \frac{5}{6}$, and $A = \frac{20}{18} = \frac{10}{9}$; then

$C = 1$, $B = \frac{5}{6}$, $A = \frac{10}{9} = \frac{6}{6}, \frac{5}{6}, \frac{10}{9}$; coef. $= \frac{18..15..20}{53}$, as before.

When the required parts are given in relation as greater or less, one than another, by a given independent quantity of their kind, to find their algebraic relation, one to another, and their values in the denomination of the dividend.

Rule.—Let x represent the first or initial algebraic part in all cases, and let x plus the given number greater than that part, or x minus the given number less than that part, as the case may be, represent the next algebraic part, and so on for all the parts involved in the proposition; then find the value of x in the denomination of the dividend for the first part, and to that part add, or from it subtract, as the case may demand, the given difference between that part and the next for the second part, and so on for all the parts required.

Prop. 69.—It is required to divide \$5000 into three parts, such that the second shall be greater than the first by \$600, and the third less than the first by \$400. The algebraic relations of the parts, and the parts in the denomination of the dividend, are demanded.

$x, x+600, x-400$, the algebraic relations of the parts, in their order, that, collectively taken, are equal to the dividend; then

$x+x+600+x-400 = 3x+200 = \$5000$, and
$3x = 5000-200 = \$4800$, therefore
$x = 4800 \div 3 = 1600$, 1st part.
$1600+600 = 2200$, 2d part.
$1600-400 = 1200$, 3d part.

Prop. 70.—How shall 50 acres of land be parcelled between A, B, and C, so that B's share shall contain 6 acres more than As, and C's 14 acres less than B's?

The sum of x, $x+6$, $x+6-14 = 3x-2 = 50$, therefore
$3x = 52$, and $x = 17\frac{1}{3}$ acres, A's share,
$17\frac{1}{3}+6 = 23\frac{1}{3}$ acres, B's share,
$23\frac{1}{3}-14 = 9\frac{1}{3}$ acres, C's share. } *Ans.*

When of the three parts that make up the whole, one of them and the difference of the other two are given, to find the remaining parts, or coefficients of the dividend.

Let 1 represent the sum of the parts, A, B, and C the parts, and d the given difference; then
$1-A=B+C$, and $(B+C+d)\div 2=B;\ B-d=C.$

PROP. 71.—A, B and C own a ship; A owns $\frac{7}{25}$ of it, and B owns $\frac{4}{15}$ of it more than C. What shares of it do B and C own respectively?
$1-\frac{7}{25}=\frac{18}{25}$, and $(\frac{18}{25}+\frac{4}{15})\div 2=\frac{370}{750}=\frac{37}{75}=$ B's share, and $\frac{37}{75}-\frac{4}{15}$ $=\frac{17}{75}=$ C's share. *Ans.*

When the quantities of two or more articles of unlike rates of value per unit of measure are separately given, and the relations of the rates as greater or less, one than another, by a stipulated independent rate of the same kind are given, and the sum of the products of the quantities by their rates is given, to find the rates and the values of the quantities.

Let A, B, C, &c., represent the given quantities:
Let b represent the given rate that B is greater or less than A;
Let c represent the given rate that C is greater or less than A or B, &c.;
Let D represent the given dividend or sum of the products of the quantities by their rates;
Let r represent the rate per unit of measure of lot A, lot B, lot C, &c.

When two quantities only are involved in the proposition, and A is the lesser.

r of lot $A=\dfrac{D-Bb}{A+C}$, and r of lot $B=\dfrac{D+Ab}{A+B}$; therefore

$\dfrac{(D-Bb)A}{A+B}=$ value of lot A, or A's share of the dividend; and

$\dfrac{(D+Ab)B}{A+B}=$ value of lot B, or B's share of the dividend. Also,

If we let x represent the rate of lot A or the lower of the two rates, $x(A+B)+bB=D$, $x(A+B)=D-bB$, and $x=\dfrac{D-bB}{A+B}$.

$$x+b=\frac{D-bB+(A+B)b}{A+B}=\frac{D+Ab}{A+B}.$$

PROP. 72.—Two lots of cheese, one of three hundred pounds (A's lot), and the other of 400 pounds (B's lot), were sold together by consent of parties for $126, and it was agreed between the parties that in dividing the money B's cheese should be reckoned worth

2 cents a pound more than A's. How much of the money received for the cheese belonged to A, and how much to B?

$\frac{126 - 400 \times .02}{300+400} = \frac{118}{700} = 16\frac{6}{7}$ cents a pound, the rate of A's cheese, and

$.16\frac{6}{7} \times 300 = \$50\frac{4}{7}$, A's share of the money.

$\frac{126 + 300 \times .02}{700} = \frac{132}{700} = 18\frac{6}{7}$ cents a pound, the rate of B's cheese, and

$.18\frac{6}{7} \times 400 = \$75\frac{3}{7}$, B's share of the money.

PROP. 73.—A lumber dealer sells 8000 feet of boards consigned by A, and 4500 feet consigned by B, together for \$496, and wishes to enter the sales to the credit of the two accounts, calling B's lot worth \$2⅝ per *M* less than A's. At what prices per *M* must he enter the two lots?

$x(A+B)-Ab = 496-8 \times 2\frac{5}{8} = \$475$, and $x = 475 \div 12.5 =$ \$38 per *M*, B's lot.
$38+2\frac{5}{8}$ or $(475+12.5 \times 2\frac{5}{8}) \div 12.5 = \$40.625$ per *M*, A's lot. } *Ans.*

When more than two quantities are involved in the proposition, and x is taken to represent the rate of one of them.

$x(A+B+C, \&c.) \pm Bb \pm Cc \pm Dd, \&c. = D.$

PROP. 74.—Three lots of wool, A's of 380 pounds, B's of 450 pounds, worth 4½ cents a pound more than A's, and C's of 620 pounds, worth 2 cents a pound less than B's, were sold together for \$920.25. What was A's, what B's, and what C's share of the money, predicating the division upon the given conditions?

$$D = x(A+B+C)+Bb+C(b-c);$$
$$x = [D-Bb-C(b-c)] \div (A+B)+C); \text{ then}$$

$x = \frac{920.25-37.75}{380+450+620} = \frac{884.50}{1450} = 61$ cents, the rate per pound of A's wool; therefore

$$\left.\begin{aligned} .61 \times 380 &= \$231.80, \text{ A's share,} \\ (.61+.045) \times 450 &= 294.75, \text{ B's share,} \\ (.61+.045-.02) \times 620 &= 393.70, \text{ C's share.} \end{aligned}\right\} \textit{Ans.}$$

The individual relations of a plurality of agents to a proposed end given, to find their combined relation to the same end, when time is an element in the calculations.

GENERAL RULE.—Reduce the given relations to simple fractions (if they are not already so reduced) without changing their values, and invert them; then take the denominator of their sum for the numerator of their combined relation, and take the numerator of their sum for the denominator of their combined relation.

PROP. 75.—A can do a given piece of work in 5 days, and B can do it in 7. In what time can both together do it?

Since A can do it in 5 days he can do $\frac{1}{5}$ of it in 1 day, and since B can do it in 7 days he can do $\frac{1}{7}$ of it in 1 day; both together, therefore, can do $\frac{1}{5}+\frac{1}{7} = \frac{12}{35}$ of it in 1 day, and together they can do the whole in $\frac{35}{12} = 2\frac{11}{12}$ days. *Ans.*

PROP. 76.—A reservoir has two receiving pipes and one discharging pipe; by one of the pipes it can be filled in 108 hours, by the other in 168 hours, and by the discharging pipe it can be emptied, when full, in 84 hours. In what time can it be filled when the three pipes are acting together, supposing the velocity of the water through the pipes to be uniform?

$\frac{1}{108}+\frac{1}{168}-\frac{1}{84} = \frac{5040}{1524096}$, and $\frac{1524096}{5040} = 302.4$ hours. *Ans.*

PROP. 77.—A can do a given piece of work in 12 days, B can do 3 times as much in 27 days, and C can do 4 times as much in 35 days. In what time can they do the proposed work, working at it together?

$\frac{1}{12}+\frac{3}{27}+\frac{4}{35} = \frac{3501}{11340}$, and $\frac{11340}{3501} = 3.239+$ days. *Ans.* Or, since A can do $\frac{1}{12}$ of it in 1 day, B $\frac{1}{9}$ of it in 1 day, and C $\frac{1}{8\frac{3}{4}}$ of it in 1 day, they can collectively do $\frac{1}{12}+\frac{1}{9}+\frac{1}{8\frac{3}{4}} = \frac{291\frac{3}{4}}{945}$ of it in 1 day, and they can do the whole in $\frac{945}{291\frac{3}{4}} = 3.239+$ days, as before.

PROP. 78.—If A can accomplish a given task in $2\frac{1}{2}$ days, B in $3\frac{1}{4}$ days, and C in 5 days; in what time can they together accomplish it?

$2\frac{1}{2}, 3\frac{1}{4}, 5 = \frac{5}{2}, \frac{13}{4}, \frac{5}{1}$, inverted $= \frac{2}{5}, \frac{4}{13}, \frac{1}{5}$, and $\frac{2}{5}+\frac{4}{13}+\frac{1}{5} = \frac{295}{325}$, the relation that one day's labor of all three bears to the whole task, and $\frac{295}{325}$ inverted $= \frac{325}{295} = 1\frac{6}{59}$ days, the time required by all to accomplish it. *Ans.*

PROP. 79.—What number diminished by the difference between $\frac{4}{5}$ and $\frac{1}{3}$ of itself will equal 128?

Let x represent the number; then

$$x-\frac{4x}{5}\sim\frac{1x}{3} = x-\frac{7x}{15} = \frac{8x}{15} = 128, \text{ and } x = \frac{128\times 15}{8} = 240. \quad \textit{Ans.}$$

PROP. 80.—There are two numbers whose difference is $117\frac{1}{2}$, and one of them is $\frac{2}{7}$ less than the other, what are the numbers?

Let x represent the greater, then $x-\frac{2x}{7}$ will represent the less, and $\frac{2x}{7}=117\frac{1}{2}$; therefore $x=411\frac{1}{4}$, and $\frac{5x}{7} = \frac{411\frac{1}{4}\times 5}{7} = 293\frac{3}{4}$. *Ans.*

PROP. 81.—A merchant lost $\frac{3}{8}$ of the money he invested in trade, and then gained \$1250, when he had \$3680. How much did he lose?

Let x represent the sum invested; then

$x-\frac{3x}{8}+1250=\$3680$, $\frac{5x}{8}=3680-1250=2430$, and $x=\frac{2430\times 8}{5}$ $=\$3888$, therefore $3888\times 3\div 8=\$1458$ loss. *Ans.*

PROP. 82.—A merchant owning $\frac{5}{7}$ of a ship sold $\frac{3}{5}$ of his share for \$12000. How much at that rate is the whole ship worth?

Since $\frac{3}{5}$ of $\frac{5}{7}$ $(\frac{15}{35})$ of the ship is worth \$12000, $\frac{35}{35}$ of it must be worth $\frac{35}{15}=2\frac{1}{3}$ times more than \$12000 = \$12000 × 35 ÷ 15 = \$28000. *Ans.*

PROP. 83.—If A can do a certain job of work in $16\frac{1}{2}$ days, what part of it at that rate can he do in $14\frac{3}{4}$ days?

Since he can do $\frac{1}{16\frac{1}{2}}$ of it in 1 day, he can do $\frac{14\frac{3}{4}}{16\frac{1}{2}}$ of it in $14\frac{3}{4}$ days $=\frac{59}{66}$ of it in $14\frac{3}{4}$ days. *Ans.*

PROP. 84.—A certain quantity is to be divided into three parts such that the 1st shall be $16\frac{1}{2}$, the 2d equal to $\frac{2}{5}$ of the whole, and the 3d as much as the other two. What are the parts?

Let x represent the sum of the parts; then, by the stipulations,

$\frac{2x}{5}+16\frac{1}{2}=\frac{1}{2}x=\frac{4x+165}{10}$, therefore $5x=4x+165$, and $x=165$, then

$$\left.\begin{array}{ll}\text{1st part} & =16\frac{1}{2}\\ \text{2d part}=165\times 2\div 5 & =66\\ \text{3d part}=16\frac{1}{2}+66 & =82\frac{1}{2}\end{array}\right\}\textit{Ans.}$$

PROP. 85.—A certain estate is to be divided into three parts as follows, viz: the 1st is to be equal to $\frac{3}{7}$ of it, the 2d equal to $\frac{4}{5}$ of the remainder, and the difference between the 1st and 2d is to be \$450. What are the parts?

Let x represent the whole estate; then

$\frac{4}{5}\left(x-\frac{3x}{7}\right)=\frac{16x}{35}$, and $\frac{16x}{35}-\frac{3x}{7}=450=\frac{7x}{245}$, therefore $\frac{450\times 245}{7}$ $=x=\$15750$, then

$$\left.\begin{array}{lr}\text{1st part}= & 15750\times 3\div 7=\$6750,\\ \text{2d part}= & (15705-6750)\times 4\div 5=\ 7200,\\ \text{3d part}= & (15750-(6750+7200)=\ 1800.\end{array}\right\}\textit{Ans.}$$

PROP. 86.—What is the least number of pounds of tea that will fill canisters holding either 2, 3, 4 or 5 pounds?

The number required must be equal to the least common multiple of the given capacities of the vessels; then

$$\begin{array}{r|l}2 & 2\ .\ .\ 3\ .\ .\ 4\ .\ .\ 5\\ & 1\ .\ .\ 3\ .\ .\ 2\ .\ .\ 5=60.\ \ \textit{Ans.}\end{array}$$

PROP. 87.—What is the least number of gallons of liquor that will fill bottles containing either $\frac{2}{3}$, $\frac{4}{7}$, $\frac{3}{4}$ or $\frac{8}{13}$ of a gallon?

$$\begin{array}{r|l} 2 & 2\,..\,4\,..\,3\,..\,8 \\ 2 & 1\,..\,2\,..\,3\,..\,4 \\ & 1\,..\,1\,..\,3\,..\,2 = 24. \quad \textit{Ans.} \end{array}$$

PROP. 88.—How many bottles of either size, viz: $\frac{2}{3}$, $\frac{4}{7}$, $\frac{3}{4}$ or $\frac{8}{13}$ gallons capacity will be required to hold 24 gallons?

$24 \times \frac{3}{2} = 36$, $24 \times \frac{7}{4} = 42$, $24 \times \frac{4}{3} = 32$, $24 \times \frac{13}{8} = 39$. *Ans.*

PROP. 89.—From a cask containing 45 gallons of 96 per cent. alcohol, how much must be drawn out and replaced with water, to make 45 gallons of 90 per cent. alcohol?

$(96-90) \times 45 \div 96 = 2\frac{13}{16}$ gallons. *Ans.*

Proof.—$(45-2\frac{13}{16}) \times 96 \div 45 = 90$. See page 117.

PROP. 90.—Three persons start together from the same point in the circumference of a circle and travel around it in the same direction; one makes $\frac{3}{4}$ of a revolution in a day, another $\frac{6}{15}$, and the third $\frac{24}{75}$; in how many days will they be together again at the point of starting?

RULE.—Reduce the given relations to a common denominator, and divide that denominator by the greatest common measure of the numerators; or reduce the given relations to their lowest common numerator and divide the least common multiple of the denominators by that numerator; or reduce the given relations to their lowest terms, and divide the least common multiple of the denominators by the greatest common measure of the numerators.

EXAMPLE.—$\frac{3}{4}\,..\,\frac{6}{15}\,..\,\frac{24}{75} = \frac{3375\,..\,1800\,..\,1440}{4500}$ and

$$\begin{array}{r|l} 45 & 3375\,..\,1800\,..\,1440 \\ & \quad 75 \quad\;\; 40 \quad\;\; 32 \end{array} = \frac{4500}{45} = 100 \text{ days. } \textit{Ans.}$$

Or $\frac{3}{4}\,..\,\frac{6}{15}\,..\,\frac{24}{75} = \frac{24}{32}\,..\,\frac{24}{60}\,..\,\frac{24}{75}$, and

$$\begin{array}{r|l} 4 & 32\,..\,60\,..\,75 \\ 15 & 8\,..\,15\,..\,75 \\ & 8\,..\,1\,..\,5 = 2400 \div 24 = \end{array}$$

100 days. *Ans.*

Or $\frac{3}{4}\,..\,\frac{6}{15}\,..\,\frac{24}{75} = \frac{3}{4}\,..\,\frac{2}{5}\,..\,\frac{8}{25}$, and

$$\frac{\begin{array}{r|l} 5 & 4\,..\,5\,..\,25 \\ & 4\,..\,1\,..\,5 = 100 \end{array}}{\begin{array}{r|l} 1 & 3\,..\,2\,..\,8 = \quad 1 \end{array}} = 100 \text{ days. } \textit{Ans.}$$

PROP. 91.—Two travelers start from two points which are $76\frac{1}{2}$ miles apart, and travel toward each other till they meet, when one has traveled $11\frac{3}{4}$ miles more than the other. How many miles has each traveled?

Let x represent the shorter distance; then

$x+x+11\frac{3}{4} = 76\frac{1}{2}$, $2x = 76\frac{1}{2}-11\frac{3}{4} = 64\frac{3}{4}$, and $\left.\begin{array}{r} x = 32\frac{3}{8} \\ x+11\frac{3}{4} = 44\frac{1}{8} \end{array}\right\}$ *Ans.*

PROP. 92.—There are two numbers whose sum is $76\frac{1}{2}$, and whose difference is $11\frac{3}{4}$. What are the numbers?

Let x represent the greater number; then

$$\left.\begin{array}{r} x+x-11\frac{3}{4}=76\frac{1}{2},\ 2x=88\frac{1}{4},\ x=44\frac{1}{8} \\ 44\frac{1}{8}-11\frac{3}{4}=32\frac{3}{8} \end{array}\right\}\ \textit{Ans.}$$

See PROBLEMS, page 169.

PROP. 93.—A man traveled in a straight line 9 hours, consecutively, from home at the rate of 8 miles an hour, and toward home (on his return) at the rate of 3 miles an hour. How far from home did he proceed?

GENERAL RULE.—Divide the product of the different speeds per hour by their sum, and multiply the quotient by the whole time consumed. Or, let x represent half the number of miles traveled, or the number of miles traveled outward; then

$\frac{x}{8}$ = time consumed in traveling outward, and

$\frac{x}{3}$ = time consumed in traveling back; therefore

$\frac{x}{8}+\frac{x}{3}$ = whole time consumed = 9 hours, and

$11x=216,\ x=216\div 11=19\frac{7}{11}$ miles. *Ans.*

PROP. 94.—A starts on a journey and travels at the rate of 4 miles an hour. With what speed must B travel, who starts to overtake A 3 hours later, that he may overtake him in 5 hours?

$$v'=\frac{v(t+t')}{t'}=\frac{vt+vt'}{t'}=\left.\begin{array}{r}4\times 3=12\\ 4\times 5=20\end{array}\right\}\div 5=6\frac{2}{5}\text{ miles per hour.}\quad \textit{Ans.}$$

PROP. 95.—A horse started suddenly at a speed of 16 miles an hour and travelled at a uniformly diminishing speed for 3 hours, when his speed had become at the rate of 7 miles an hour. How many miles did he travel?

RULE.—Multiply the square root of the product of the given extremes of speed into the given time, for the answer; then

$m=t\sqrt{(vv')}=3\times\sqrt{(16\times 7)}=31.749$ miles. *Ans.*

PROP. 96.—What are the eight special weights with which, upon a pair of grocer's scales, any number of ounces from 1 to 3280, inclusive, may be weighed?

1, 3, 9, 27, 81, 243, 729, 2187 oz. *Ans.*

PROP. 97.—What sum in ready money may I pay for a mortgage of $3000, drawing interest at 7 per cent., and payable in sums of $1000 and interest at the close of each year from the present time, till paid, in order that I may realize 10 per cent. interest on the investment?

$[(p+Pr)\times(1+r)^{(n-1)}+p+(P-p)r\times(1+r)^{(n-2)}$ &c.$]\div(1-a)^n=$ present worth, allowing compound discount, and compound interest on the payments, then

Amount (principal and interest) three years hence of

1st payment $=(1000+3000\times.07)\times(1.07)^2=\$1385.33$
2d payment $=(1000+2000\times.07)\times1.07 \quad = \quad 1219.80$
3d payment $=(1000+1000\times.07)\times1 \quad = \quad 1070.00$

$(1.10)^3=1.331)\ 3675.13=$
$\$2761.18.$ *Ans.*

$(p+Pr)\times[1+r(n-1)]+[p+(P-p)r]\times[1+r(n-2)]$ &c.$\div(1+rn)$ $=$ present worth, allowing simple discount, and simple interest on the payments; then

Amount (principal and interest) three years hence of

1st payment $=(1000+3000\times.07)\times1.14=\$1379.40$
2d payment $=(1000+2000\times.07)\times1.7 \quad = \quad 1219.80$
3d payment $=(1000+1000\times.07)\times1 \quad = \quad 1070.00$

$1+3a=1.30)\ 3669.20=$
$\$2822.46.$ *Ans.*

PROP. 98. — Ganson & Co. are paying \$12,000 a year in monthly payments in advance for the use of the premises they occupy; and they wish to know how much less the rent would be per annum, were it fixed at the same price by the year, but payable quarterly, at the expiration of each quarter-year; allowing money to be worth 7 per cent. interest a year.

$$P+\frac{Pr(n+1)}{2n}=12p+\frac{pr(n+1)}{2}=\$12{,}455.00,$$ the amount, by the present manner of paying, at the close of the year; and

$$P+\frac{Prn}{2(n+1)}=12p+\frac{pr(12-n)}{2}=\$12{,}315.00,$$ the amount, by the proposed manner of paying, at the close of the year, in both of which cases n represents the number of payments to be made per annum; then

$$12455-12315=\$140.00.\quad \textit{Ans.}$$

PROP. 99.—A and B took a job in company of digging a ditch of a certain width and depth and 100 rods in length for \$100. They agreed between themselves to dig separately from the opposite ends of the ditch toward each other, and that each should do one-half of the work and receive one-half of the pay, and that, because of the difference in the soil, A's allowance per rod for digging should be to that of B's as 75 cents to \$1.25. How many rods must each dig?

$\$50\div0.75=66\frac{2}{3}$
$50\div1.25=40$

$106\frac{2}{3}$

$106\frac{2}{3} : 100 :: 66\frac{2}{3} : 62\frac{1}{2}$ rods, to be dug by A }
$106\frac{2}{3} : 100 :: 40 : 37\frac{1}{2}$ rods, to be dug by B } *Ans.*

$62\frac{1}{2} : 66\frac{2}{3} :: 75 : 80$ cents per rod, A's price for digging.
$37\frac{1}{2} : 40 :: 1.25 : \$1\frac{1}{3}$ per rod, B's price for digging.

CONFIRMATION.

$62.5 \times 0.80 = \$50$
$37.5 \times 1\frac{1}{3} = 50$

100 rods $= \$100$

$.75 : .80 :: 1.25 : 1.33\frac{1}{3}$.

PROP. 100.—A and B own a stock of goods worth $10,000, and one-third of A's interest in the stock is equal to two-fifths of B's. What is the interest of each?

$\frac{2}{5} .. \frac{1}{3} = \frac{6..5}{11}$, then

$10,000 \times 6 \div 11 = \$5454\frac{6}{11}$, A's share }
$10,000 \times 5 \div 11 = \$4545\frac{5}{11}$, B's share } *Ans.*

Or, let $x =$ A's share, then $10,000 - x$ will represent B's share, and by the conditions $\frac{(10,000-x)2}{5} = \frac{x}{3}$; therefore $\frac{20,000-2x}{5} = \frac{x}{3}$, and $60,000-6x = 5x$, then $11x = 60,000$, $x = \$5454\frac{6}{11}$, and $10,000 - 5454\frac{6}{11} = \$4545\frac{5}{11}$.

www.ingramcontent.com/pod-product-compliance
Lightning Source LLC
LaVergne TN
LVHW010256110826
845151LV00004B/1484